Lecture Notes in Economics and Mathematical Systems

Managing Editors: M. Beckmann and W. Krelle

339

Jaime Terceiro Lomba

Estimation of Dynamic Econometric Models with Errors in Variables

Springer-Verlag

Berlin Heidelberg New York London Paris Tokyo Hong Kong

Editorial Board

H. Albach M. Beckmann (Managing Editor)
P. Dhrymes G. Fandel G. Feichtinger J. Green W. Hildenbrand W. Krelle (Managing Editor) H. P. Künzi K. Ritter R. Sato U. Schittko P. Schönfeld R. Selten

Managing Editors

Prof. Dr. M. Beckmann
Brown University
Providence, RI 02912, USA

Prof. Dr. W. Krelle
Institut für Gesellschafts- und Wirtschaftswissenschaften
der Universität Bonn
Adenauerallee 24–42, D-5300 Bonn, FRG

Author

Prof. Dr. Jaime Terceiro Lomba
Facultad de Económicas, Universidad Complutense
Campus de Samosaguas, 28023 Madrid, Spain

ISBN 3-540-52358-8 Springer-Verlag Berlin Heidelberg New York
ISBN 0-387-52358-8 Springer-Verlag New York Berlin Heidelberg

This work is subject to copyright. All rights are reserved, whether the whole or part of the material is concerned, specifically the rights of translation, reprinting, re-use of illustrations, recitation, broadcasting, reproduction on microfilms or in other ways, and storage in data banks. Duplication of this publication or parts thereof is only permitted under the provisions of the German Copyright Law of September 9, 1965, in its version of June 24, 1985, and a copyright fee must always be paid. Violations fall under the prosecution act of the German Copyright Law.

© Springer-Verlag Berlin Heidelberg 1990
Printed in Germany

Printing and binding: Druckhaus Beltz, Hemsbach/Bergstr.
2142/3140-543210 – Printed on acid-free paper

A mis padres
J.T.L.

PREFACE

Most of the results contained in this work were obtained in 1975, 1976 and 1977. A preliminary version of the computer program used was also developed at that time.

These notes were employed for the first time in the doctoral course that I taught in academic year 1984-85. Blanca Fernández Cassoni, a student in that course, has contributed to the empirical results shown here.

A preliminary form of the text was sent to the Economics Editor of "Lecture Notes in Economics and Mathematical Systems" in October, 1986. This work is a slightly modified version of it and includes reference to some of the recent publications in the field of errors in variables. Professor Arthur Treadway read most of this text and made useful comments on it.

Jaime Terceiro Lomba

TABLE OF CONTENTS

1. Introduction 1
2. Formulation of Econometric Models in State-Space 5
   - 2.1. Structural Form, Reduced Form and State-Space Form 5
   - 2.2. Additional Remarks 12
3. Formulation of Econometric Models with Measurement Errors 17
   - 3.1. Model of the Exogenous Variables 17
   - 3.2. State-Space Formulation 20
4. Estimation of Econometric Models with Measurement Errors 24
   - 4.1. Evaluation of the Likelihood Function 26
   - 4.2. Maximization of the Likelihood Function 29
   - 4.3. Initial Conditions 35
   - 4.4. Gradient Methods and Identification 37
   - 4.5. Asymptotic Properties 41
   - 4.6. Numerical Considerations 43
   - 4.7. Model Verification 45
5. Extensions of the Analysis 49
   - 5.1. Missing Observations and Contemporaneous Aggregation 49
   - 5.2. Temporal Aggregation 51
   - 5.3. Correlated Measurement Errors 53
6. Numerical Results 56
7. Conclusions 68

Appendices

A. Kalman Filter and Chandrasekhar Equations 70

    A.1. Kalman Filter 70
    A.2. Chandrasekhar Equations 73

B. Calculation of the Gradient 80

C. Calculation of the Hessian 83

D. Calculation of the Information Matrix 85

E. Estimation of the Initial Conditions 90

F. Solution of the Lyapunov and Riccati Equations 94

    F.1. Lyapunov Equation 94
    F.2. Riccati Equation 97

G. Fixed-Interval Smoothing Algorithm 101

References 105

Author Index 111

Subject Index 113

CHAPTER 1

INTRODUCTION

As is well-known, econometric estimation procedures have not placed much emphasis on the problem of errors in variables. What is more, the fact that the regression model with measurement errors in the independent variables is not identified may lead to the erroneous belief that the treatment of this problem in more complex situations is even less open to study.

The econometric problem of errors in variables acquires importance when models, which include theoretical or abstract variables and for which measurements are imperfect are constructed. Examples of such variables could be utility, ability or human capital.

Formulations which consider observation errors explicitly are appropriate when one is interested in the "true" system generating the observations. On the other hand, the problem may arise as a result of measurement errors in the variables. An example of this situation is Friedman's (1957) permanent income hypothesis. In some cases in empirical analysis the data may be subject to errors or the variables we measure are not really the ones we want to measure. In other situations preliminary observations are used which are frequently subject to substantial revisions.

We shall also see that situations of spurious feedback among variables and overparametrization of certain formulations can be avoided with an error in variables formulation.

In any case, the state-space formulation proposed in this work allows a general approach to a series of problems which have recently been dealt with in the econometrics literature.

This work puts forward a general procedure for the maximun likelihood estimation of dynamic econometric models with errors in both endogenous and exogenous variables. The basic idea consists in considering the endogenous and exogenous variables of the original specification as observable variables in a new state-space parametrization. Given appropriate regularity conditions, it can be shown that these maximum likelihood estimates are consistent and asymptotically normal and efficient with a covariance matrix given by the inverse of the information matrix.

Aoki (1987) has recently proposed a new approach to state-space formulations which allows one to get the specification of the model directly from a data-based procedure, and estimates the parameters by substituting sample for population moments in equations derived from state-space theoretical considerations. Unlike the maximum likelihood approach proposed in this work, the statistical properties of these estimators are not completely developed. In addition, the state-space formulation here obtained in Chapters 2 and 3 comes from structural formulations of econometric models derived from economic theory or from statistical considerations so that the matrices of the model include a priori restrictions which avoid their possible overparametrization.

A large proportion of recent results for dynamic models with errors in variables is dealt with in the survey paper by Aigner et al. (1984). The analysis of identifiability for some types of time series with errors in variables, carried out by these authors, has been substantially simplified in the work by Solo (1986). We should also point out in this context the work by Anderson and Deistler (1984) and Deistler (1986).

The basic characteristic of all of this recent literature on econometric models with errors in variables is that it only deals with aspects of identification and generally has been limited to obtaining merely consistent estimates by solving the corresponding covariance equations. However, these results are not easily applied generally to simultaneous equations models and they do not lead to efficient estimation.

The work is organized as follows. Chapter 2 deals with the formulation of dynamic econometric models in state-space. Unlike parametrizations commonly employed in the econometrics literature, the one which we use here has a state vector with minimal dimension. Chapter 3 adds to this formulation the possibility of measurement errors in both the endogenous and exogenous variables. This requires the previous formulation in state-space of the multivariate ARMA process which, in general, generates the exogenous variables.

The problem of estimating the model is undertaken in Chapter 4. This is done first by constructing the likelihood function, followed by the analytical expressions of its gradient and hessian. The exact expression of the information matrix is also obtained. The information matrix plays an important role not only in the numerical optimization algorithm used, but also, from a statistical point of view, as a basic element in the process of the validation of the model. In this chapter, reference is also made, from a practical perspective, to the problem of identification of the model through a numerical analysis of the information matrix as well as to the difficulties associated with the optimization algorithms used. Other important aspects are also examined, such as the estimation of the initial conditions of the state vector, and the use of the Chandrasekhar equations instead of the Riccati equations of the Kalman filter.

Chapter 4 together with the appendices consists of a complete analytical development of the expressions used in the problem of estimating econometric models in state-space form. Therefore, the results presented here are useful not only in relation to the problem of errors in variables that we examine, but also in relation to any other possible econometric application of state-space formulations. For example, the algorithm set out in Chapter 4 can be applied to the estimation of econometric models with rational expectations, see Hamilton (1985), to seasonal adjustment of time series, see Hausman and Watson (1985), and to the estimation of econometric models with composite moving average disturbance terms, see McDonald and Darroch (1983). It can easily be shown that these problems can be formulated as described in Chapters 2 and 3, and that they only represent special cases of the general formulation advanced here.

In Chapter 5 the proposed formulation is extended to situations with missing observations, to cases where only temporal aggregates of the variables are observed and to the case where the measurement errors are correlated.

In Chapter 6 the methodology which has been developed in previous chapters is used in the estimation of the following model:

$$y_t + \beta y_{t-1} + \gamma u_t = \varepsilon_t, \quad \varepsilon_t \sim N(0,\sigma_\varepsilon^2)$$

$$y_t^* = y_t + v_{yt}, \quad v_{yt} \sim N(0,\sigma_{v_y}^2)$$

$$u_t^* = u_t + v_{ut}, \quad v_{ut} \sim N(0,\sigma_{v_u}^2)$$

Numerical results are obtained for different levels of error in the variables, given by $\sigma^2_{v_y}$ and $\sigma^2_{v_u}$, and for different sample sizes. These are given under various assumptions about the correlation structure of the exogenous variables.

Aigner et al. (1984) analyze this model and it is precisely in this context that these authors affirm that there are no procedures available to estimate this type of models satisfactorily.

On the basis of the numerical results shown in Chapter 6, we believe that the procedures suggested in this work cover the existing gap in current literature on the estimation of econometric models with measurement errors and missing observations in both the endogenous and exogenous variables.

Finally, Chapter 7 draws conclusions and summarizes the contribution made by this work.

CHAPTER 2

FORMULATION OF ECONOMETRIC MODELS IN STATE-SPACE

In this chapter we shall analyze the relationship between the structural, reduced, and state-space forms of an econometric model. We do not use the state-space formulations usually employed in the econometrics literature, but employ one that is based on the minimum dimension for the vector of state variables and which explicitly takes into account the improper nature of econometric models.

We conclude by advancing some remarks which justify the formulation we propose.

2.1. Structural Form, Reduced Form and State-Space Form.

Let us consider the following structural form of an econometric model:

$$\bar{F}(L)y_t = \bar{G}(L)u_t + \bar{\Xi}(L)\varepsilon_t \qquad (2.1)$$

where:

y_t is a vector of l endogenous variables,
u_t is a vector of r exogenous variables, and
ε_t is a vector of disturbances of l random variables, such that

$$E[\varepsilon_t] = 0, \quad E[\varepsilon_{t_1} \varepsilon'_{t_2}] = \Sigma_\varepsilon \delta_{t_1 t_2}, \quad \delta_{t_1 t_2} = \begin{cases} 0, & t_1 \neq t_2 \\ 1, & t_1 = t_2 \end{cases}$$

The polynomial matrices $\bar{F}(L)$, $\bar{G}(L)$ and $\bar{\Xi}(L)$, of dimensions (l×l), (l×r) and (l×l), can be written:

$$\bar{F}(L) = \sum_{i=0}^{p} \bar{F}_i L^i, \quad \bar{G}(L) = \sum_{i=0}^{q} \bar{G}_i L^i, \quad \bar{\Xi}(L) = \sum_{i=0}^{s} \bar{\Xi}_i L^i \qquad (2.2)$$

where L is the shift operator. For any sequence x_t:

$$L^{\pm t_1} x_t = x_{t \mp t_1}$$

The presence of $\bar{F}_0$ in the equation defining the polynomial matrix $\bar{F}(L)$ in (2.2) generates instantaneous coupling between the endogenous variables. Assuming that $\bar{F}_0$ is not singular, we obtain the reduced form:

$$F(L) y_t = G(L) u_t + \Xi(L) \varepsilon_t \qquad (2.3)$$

where:

$$F(L) = \bar{F}_0^{-1} \bar{F}(L), \quad G(L) = \bar{F}_0^{-1} \bar{G}(L), \quad \Xi(L) = \bar{F}_0^{-1} \bar{\Xi}(L) \qquad (2.4)$$

The instantaneous coupling is removed in the reduced form which therefore expresses each endogenous variable as an explicit function of lagged endogenous variables and current and lagged exogenous variables and disturbances.

If the reduced form is stable (the roots of F(L) lie outside the unit circle) the matrix F(L) can be inverted to give the final form:

$$y_t = \Pi(L) u_t + \Psi(L) \varepsilon_t \qquad (2.5)$$

with:

$$\Pi(L) = F^{-1}(L) G(L), \quad \Psi(L) = F^{-1}(L) \Xi(L) \qquad (2.6)$$

The matrices $\Pi(L)$ and $\Psi(L)$ are rational functions of L and can be expanded in a power series in L so that

$$\Pi(L) = \sum_{i=0}^{\infty} \Pi_i L^i, \quad \Psi(L) = \sum_{i=0}^{\infty} \Psi_i L^i$$

The matrices Π_i are the dynamic matrix multipliers, Π_0 is the impact matrix multiplier and $\Pi_\infty = \sum_{i=0}^{\infty} \Pi_i$ is the steady-state matrix multiplier.

In the final form (2.5) each endogenous variable is expressed as infinite distributed-lag functions of exogenous variables and disturbances.

In the state-space formulation, the model dynamics are expressed by means of a first-order system of difference equations using auxiliary variables which are called state variables. This description of linear dynamic systems has become dominant in the control literature after the pioneering work of Kalman (1960).

The state-space model which we shall use in this work is given by:

$$x_{t+1} = \Phi x_t + \Gamma u_t + E w_t \qquad (2.7)$$

$$z_t = H x_t + D u_t + C v_t \qquad (2.8)$$

where:

x_t is the vector of the state variables of dimension n,

u_t is the vector of the exogenous variables of dimension r,

z_t is the vector of the observed variables of dimension m, which does not in general coincide with the vector of endogenous variables, and

w_t and v_t are white noise processes such that:

$$E[w_t] = 0, \quad E[v_t] = 0 \qquad (2.9)$$

$$E\left\{ \begin{bmatrix} w_{t_1} \\ v_{t_1} \end{bmatrix} \begin{bmatrix} w'_{t_2} & v'_{t_2} \end{bmatrix} \right\} = \begin{bmatrix} Q & S \\ S' & R \end{bmatrix} \delta_{t_1 t_2} \qquad (2.10)$$

We will assume that the matrix Q is positive semidefinite and that the matrix R is positive definite, that is:

$Q \geq 0, \quad R > 0$

Equation (2.7) corresponds to the dynamics of the system and is called the state transition equation and (2.8) corresponds to the observation of the system and is called the observation equation.

It is generally reasonable to assume that the initial state x_0 is a random variable, with mean $\bar{x}_0$ and covariance matrix P_0, uncorrelated with w_t and v_t.

Given (2.7)-(2.8) we can define a new state vector with the similarity transformation given by

$$x_t^* = T^{-1} x_t \qquad (2.11)$$

The formulation (2.7)-(2.8) in terms of these new variables is

$$x_{t+1}^* = \Phi^* x_t^* + \Gamma^* u_t + E^* w_t \qquad (2.12)$$

$$z_t = H^* x_t^* + D u_t + C v_t \qquad (2.13)$$

where

$$\Phi^* = T^{-1} \Phi T, \quad \Gamma^* = T^{-1} \Gamma, \quad E^* = T^{-1} E, \quad H^* = HT$$

If, in the process of specifying the econometric model, we adopt the criterion that the only relevant variables are the input vectors u_t, w_t and v_t, and the output vector, z_t, we see that the models (2.7)-(2.8) and (2.12)-(2.13) are input-output equivalent. Thus, within this black-box version of the matter, the state vector x_t may appear to be a strange theoretical element defined with a certain degree of ambiguity, this because the matrix T in (2.11) may be taken as any matrix with an inverse.

This ambiguity may reveal identification problems in the model (2.7)-(2.8), because it may be impossible to distinguish between the models (2.7)-(2.8) and (2.12)-(2.13) given sample information on the variables u_t and z_t. Therefore, in order to make the general parametrization (2.7)-(2.8) identifiable, it is necessary to impose a set of restrictions on the matrices Φ, Γ, E, H, D, C, Q, R, S and on the characteristics of the vector u_t. When such restrictions are imposed, we obtain what we will call canonical parametrizations, which are identifiable formulations, see Wertz et al. (1982) and Glover and Willems (1974).

Undoubtedly, another procedure for avoiding the problems just mentioned is to eliminate the state variables with the objective of transforming (2.7)-(2.8) into the reduced form (2.3) or the final form (2.5). We now consider how to do this.

If we want to compare (2.11)-(2.12) with (2.3) or (2.5), we will have to assume that the vector of observed variables, z_t, coincides with the vector of endogenous variables, y_t. In the context of the present study, this is equivalent to assuming that both exogenous and endogenous variables are observed without error. Therefore, the state-space formulation we shall use is:

$$x^a_{t+1} = \Phi^a x^a_t + \Gamma^a u_t + E^a w^a_t \qquad (2.14)$$

$$y_t = H^a x^a_t + D^a u_t + C^a v^a_t \qquad (2.15)$$

where the white noises processes w^a_t, v^a_t are such that:

$$E\left[w^a_t\right] = 0, \quad E\left[v^a_t\right] = 0 \qquad (2.16)$$

$$E\left\{\begin{bmatrix} w^a_{t_1} \\ v^a_{t_1} \end{bmatrix} \begin{bmatrix} (w^a_{t_2})' & (v^a_{t_2})' \end{bmatrix}\right\} = \begin{bmatrix} Q^a & S^a \\ (S^a)' & R^a \end{bmatrix} \delta_{t_1 t_2} \qquad (2.17)$$

with $Q^a \geq 0$ and $R^a > 0$.

Using the operator L in (2.14) we obtain:

$$x^a_t = L(I - L\Phi^a)^{-1}\Gamma^a u_t + L(I - L\Phi^a)^{-1} E^a w^a_t \qquad (2.18)$$

and substituting this in (2.15) gives:

$$y_t = \left[H^a L(I - L\Phi^a)^{-1}\Gamma^a + D^a\right]u_t + H^a L(I - L\Phi^a)^{-1}E^a w_t^a + C^a v_t^a \quad (2.19)$$

In this way we have eliminated the state variables and we can compare (2.19), obtained from the state-space formulation, with the reduced form (2.3) and the final form (2.5). Comparing these expressions, we can affirm that (2.19) and the reduced form (2.3) coincide if:

i) $\quad H^a L(I - L\Phi^a)^{-1}\Gamma^a + D^a = F(L)^{-1} G(L)$ \quad (2.20)

ii) $\quad H^a L(I - L\Phi^a)^{-1} E^a w_t^a + C^a v_t^a$ has the same autocorrelation function as $F(L)^{-1} \Xi(L)\varepsilon_t$ \quad (2.21)

Analogously, the form (2.19) coincides with the final form (2.5) if:

i') $\quad H^a L(I - L\Phi^a)^{-1}\Gamma^a + D^a = \Pi(L)$ \quad (2.22)

ii') $\quad H^a L(I - L\Phi^a)^{-1} E^a w_t^a + C^a v_t^a$ has the same autocorrelation function as $\Psi(L)\varepsilon_t$ \quad (2.23)

Conditions (2.22) and (2.23) reveal that the matrices Φ^a, Γ^a, E^a, H^a, D^a, C^a, Q^a, R^a, S^a of the state-space form determine the matrices Π_i, Ψ_i, Σ_ε of the final form uniquely. However, the matrices of the final form do not determine those of the state-space form uniquely. Analogously, conditions (2.20) and (2.21) indicate that the state-space form does not determine the reduced form uniquely and viceversa.

The problem of obtaining the matrices Φ^a, Γ^a, E^a, H^a, D^a, C^a, Q^a, R^a, S^a from $F(L)$, $G(L)$, $\Xi(L)$, Σ_ε has been analyzed in the control literature, where it is known as the stochastic realization problem.

As we have shown, it is possible to obtain an infinite number of realizations in state-space from the traditional forms of econometric models. Nevertheless, the only interesting ones are those with a minimum number of state variables; we call these minimal realizations. Rosenbrock (1970) showed that a state-space realization is minimal if it is observable and reachable. This is equivalent to requiring the observability condition:

$$\text{rank}\left[(H^a)' \; (\Phi^a)'(H^a)' \; (\Phi^a)'^2(H^a)' \; \ldots \; (\Phi^a)'^{n-1}(H^a)'\right] = n \qquad (2.24)$$

and the reachability conditions:

$$\text{rank}\left[\Gamma^a \; \Phi^a\Gamma^a \; (\Phi^a)^2\Gamma^a \; \ldots \; (\Phi^a)^{n-1}\Gamma^a\right] = n$$

$$\text{rank}\left[E^a \; \Phi^a E^a \; (\Phi^a)^2 E^a \; \ldots \; (\Phi^a)^{n-1}E^a\right] = n \qquad (2.25)$$

That is, the rank of the observability and reachability matrices must equal the dimension of the state vector.

In this study we use state-space realizations obtained from the reduced form. It is simple to demonstrate, see Aoki (1976), that the reduced form (2.3) can be written in the state-space form (2.14)-(2.15) defining:

$$\Phi^a = \begin{bmatrix} -F_1 & I & 0 & \cdots & 0 \\ -F_2 & 0 & I & \cdots & 0 \\ \cdot & \cdot & \cdot & \cdots & \cdot \\ \cdot & \cdot & \cdot & \cdots & \cdot \\ \cdot & \cdot & \cdot & \cdots & \cdot \\ -F_{k-1} & 0 & 0 & \cdots & I \\ -F_k & 0 & 0 & \cdots & 0 \end{bmatrix},$$

$$\Gamma^a = \begin{bmatrix} G_1 - F_1 G_0 \\ G_2 - F_2 G_0 \\ \cdot \\ \cdot \\ \cdot \\ G_k - F_k G_0 \end{bmatrix}, \quad E^a = \begin{bmatrix} \Xi_1 - F_1 \Xi_0 \\ \Xi_2 - F_2 \Xi_0 \\ \cdot \\ \cdot \\ \cdot \\ \Xi_k - F_k \Xi_0 \end{bmatrix} \qquad (2.26)$$

$$H^a = \begin{bmatrix} I & 0 & \cdots & 0 \end{bmatrix}, \quad D^a = G_0, \quad C^a = \Xi_0 \qquad (2.27)$$

in such a way that the dimension of the state vector is given by $k = \max(p,q,s)$.

The white noise processes in (2.14) and (2.15) are given by:

$$w_t^a = v_t^a = \varepsilon_t \qquad (2.28)$$

and with the notation of (2.16) and (2.17) their covariance matrices are:

$$Q^a = R^a = S^a = \Sigma_\varepsilon \qquad (2.29)$$

The representation defined by (2.26)-(2.29) is minimal, i.e., no representation exists with a state of lower dimension. It is easy to show that this representation satisfies conditions (2.24)-(2.25).

2.2. Additional Remarks.

As we see, two of the basic characteristics of the state-space formulation (2.7)-(2.8) are the consideration of the state vector, which is not present in the traditional formulations of econometric models given by (2.1), (2.3) and (2.5), and the presence of two white noise terms w_t and v_t instead of the single term ε_t.

The state vector of formulation (2.7)-(2.8) can be considered as the minimal information necessary to uniquely determine the evolution of the system given the future values of exogenous variables and perturbation terms beginning from a known initial value for the state variables. In practice the perturbation terms are characterized by their probability density functions. Thus the state-space formulation permits the determination of the probability density function for the state vector for successive moments in time.

The consideration of two noise terms instead of only one offers undoubted advantages in certain econometric applications. The separate estimation of Q and R supplies information that would be lost if only a combination of them, Σ_ε, were estimated. This is evident in the estimation of models with errors in variables which we will treat in detail in later chapters. But it is also evident in other kinds of formulations such as, for example, those involving the estimation of varying and random coefficients, see Cooley and Prescott (1976) and Chow (1984).

The consideration of two noise terms also offers evident advantages from the viewpoint of the parsimonious formulation of econometric models, following the approach (not to use unnecessarily many parameters) suggested by Box and Jenkins (1976).

For example, if the variable y_t follows an univariate autoregressive process of order $\bar{p}$, AR($\bar{p}$), given by

$$\bar{\phi}_{\bar{p}}(L) y_t = w_t \qquad (2.30)$$

with

$$\bar{\phi}_{\bar{p}}(L) = 1 - \bar{\phi}_1 L - \ldots - \bar{\phi}_{\bar{p}} L^{\bar{p}}$$

$$E[w_t] = 0, \quad E[w_{t_1} w_{t_2}] = Q\delta_{t_1 t_2},$$

and we observe y_t with error such that

$$z_t = y_t + v_t \qquad (2.31)$$

with

$$E[v_t] = 0, \quad E[v_{t_1} v_{t_2}] = R\delta_{t_1 t_2}, \quad E[v_t w_t] = 0$$

it is easy to show, see Granger and Morris (1976), that the observed variable, z_t, follows a univariate mixed autoregressive-moving average process of order $(\bar{p},\bar{p})$, ARMA($\bar{p},\bar{p}$), given by

$$\bar{\phi}_{\bar{p}}(L) z_t = \bar{\theta}_{\bar{p}}(L) \varepsilon_t \qquad (2.32)$$

with

$$\bar{\theta}_{\bar{p}}(L) = 1 - \bar{\theta}_1 L - \ldots - \bar{\theta}_{\bar{p}} L^{\bar{p}}$$

$$E[\varepsilon_t] = 0, \quad E[\varepsilon_{t_1} \varepsilon_{t_2}] = \Sigma_\varepsilon \delta_{t_1 t_2}$$

Thus, the formulation of the process z_t requires the estimation of $(2\bar{p}+1)$ parameters $(\bar{\phi}_1, \ldots, \bar{\phi}_p, \bar{\theta}_1, \ldots, \bar{\theta}_p, \Sigma_\varepsilon)$ while the formulation of the original process (2.30) and the observation equation (2.31) only requires the estimation of $(\bar{p}+2)$ parameters $(\bar{\phi}_1, \ldots, \bar{\phi}_p, Q, R)$.

The state-space formulation is particulary useful both when the state vector has some economic interpretation and in problems of estimation. An example of the first situation is the formulation of varying and random coefficient models. A stochastic process x_t is said to be Markov if

$$p(x_{t+j}|x_1, \ldots, x_t) \equiv p(x_{t+j}|x_t)$$

where $p(x_{t+j}|\cdot)$ is the conditional probability density of the state vector x_t. The state process x_t given by (2.7) is Markov, we shall see in Chapter 4 that this is a key property for problems of estimation.

It is interesting to observe the difference between the formulation given by (2.26)-(2.29) and those used by other authors in the econometrics literature. For example, Chow (1975, 1981) uses nonminimal dimensional models, while Harvey (1981a, 1981b) employs formulations which, in some cases, have higher dimensions than the minimal.

Unlike what occurs in physical models, there is in economic models a contemporaneous relationship between inputs and outputs due to the relatively large sampling intervals which are used in their estimation. Models which reflect this type of instantaneous causality are called improper. Models are called proper when there is no contemporaneous relationship. These properties are reflected in the previous formulations in the following way. A model will be proper with respect to the input u_t if $\Pi_0 = 0$, and with respect to the input ε_t if $\Psi_0 = 0$. Improper models are characterized by $\Pi_0 \neq 0$ and/or $\Psi_0 \neq 0$.

The following example serves to highlight the differences. The ARMA(2,2) model

$$(1 + \bar{\phi}_1 L + \bar{\phi}_2 L^2) y_t = (1 + \bar{\theta}_1 L + \bar{\theta}_2 L^2) \varepsilon_t$$

is an improper model, and can be written, following Harvey(1981a, 1981b), in the form

$$x_{t+1} = \begin{bmatrix} -\bar{\phi}_1 & 1 & 0 \\ -\bar{\phi}_2 & 0 & 1 \\ 0 & 0 & 0 \end{bmatrix} x_t + \begin{bmatrix} 1 \\ \bar{\theta}_1 \\ \bar{\theta}_2 \end{bmatrix} \varepsilon_{t+1} \qquad (2.33)$$

$$y_t = \begin{bmatrix} 1 & 0 & 0 \end{bmatrix} x_t \qquad (2.34)$$

or alternatively, according to (2.26)-(2.29), as

$$x_{t+1} = \begin{bmatrix} -\bar{\phi}_1 & 1 \\ -\bar{\phi}_2 & 0 \end{bmatrix} x_t + \begin{bmatrix} \bar{\theta}_1 - \bar{\phi}_1 \\ \bar{\theta}_2 - \bar{\phi}_2 \end{bmatrix} \varepsilon_t \qquad (2.35)$$

$$y_t = \begin{bmatrix} 1 & 0 \end{bmatrix} x_t + \varepsilon_t \qquad (2.36)$$

Note that the parametrization (2.33)-(2.34), that is, the one where the state transition equation contains a contemporaneous relationship between the state variables and the noise term, has a state vector of higher dimension than the parametrization (2.35)-(2.36). Moreover, in the formulation (2.35)-(2.36) the system noise and the observation noise are correlated.

Throughout the sequel, we will always use state-space formulations of the kind (2.26)-(2.29), that is, we will explicitly take into account the improper characteristic of econometric models. The advantages of using these formulations can be summarized as follows:

- The state variable vector has minimal dimension. This implies computational advantages.

- The noise term in the observation equation tends to avoid numerical problems resulting from the possible singularity of the covariance matrix which is propagated in the Kalman filter because a linear combination of the state variables is observed without error.

- It permits the direct application of the Chandrasekhar equations to which we will refer in Chapter 4 and Appendix A.

- In the case of small samples, frequent in econometrics, the estimates obtained in the process of maximization of the likelihood function can depend entirely upon the initial conditions of the state vector. It is the estimation of these initial conditions from sampling information that we deal with in Appendix E. Nevertheless, it is clear that when the state vector has minimal dimension the estimation of initial conditions will be more efficient than when the dimension is non-minimal, and also situations where initial conditions are not identified can be avoided more easily.

Finally, it is worth arguing in favour of the representation we propose since it is the one most frequently used in the theory of linear systems, which is rather developed, and therefore its results are applicable to these formulations.

Note that in general the unknown elements of the matrices of the state-space formulation are nonlinear functions of the basic parameters of the econometric model which we shall represent by the vector θ of p components. Thus the estimation procedure we shall use will include these restrictions on the components of θ, as well as any other following from the theoretical or empirical properties which can be imposed on the model, such as its long term behaviour (characterized by the value of $\Pi_\infty = H(I - \Phi)^{-1}\Gamma + D)$ or stability conditions. We shall only assume that the elements of the matrices of the state-space formulation are differentiable with respect to θ.

CHAPTER 3

FORMULATION OF ECONOMETRIC MODELS WITH MEASUREMENT ERRORS

In the last chapter we obtained a state-space parametrization for the dynamic econometric model given by (2.3).

In general, both endogenous and exogenous variables in model (2.3) will be observed with measurement errors. That is to say:

$$y_t^* = y_t + v_{yt} \tag{3.1}$$

$$u_t^* = u_t + v_{ut} \tag{3.2}$$

where v_{yt} and v_{ut} are the observation errors of y_t and u_t, respectively. Let us assume that both processes are white noise and fulfil the following:

$$E[v_{yt}] = 0, \quad E[v_{ut}] = 0 \tag{3.3}$$

$$E\left\{\begin{bmatrix} v_{yt_1} \\ v_{ut_1} \end{bmatrix} \begin{bmatrix} v'_{yt_2} & v'_{ut_2} \end{bmatrix}\right\} = \begin{bmatrix} \Sigma_y & 0 \\ 0 & \Sigma_u \end{bmatrix} \delta_{t_1 t_2} \tag{3.4}$$

What is required before formulating the econometric model with measurement errors in the variables is a model of the vector of exogenous variables.

3.1. Model of the Exogenous Variables.

The structural form (2.1) of the econometric model can be obtained from a multivariate ARMA model for the $[y_t' \quad u_t']'$ vector to which certain restrictions apply, see Zellner and Palm (1974). In effect, we can write this ARMA model as follows:

$$\begin{bmatrix} \phi^{11}(L) & \phi^{12}(L) \\ \phi^{21}(L) & \phi^{22}(L) \end{bmatrix} \begin{bmatrix} y_t \\ u_t \end{bmatrix} = \begin{bmatrix} \theta^{11}(L) & \theta^{12}(L) \\ \theta^{21}(L) & \theta^{22}(L) \end{bmatrix} \begin{bmatrix} \varepsilon_t \\ a_t \end{bmatrix} \quad (3.5)$$

If in this equation we impose the following restrictions:

$$\phi^{21}(L) = 0 \quad (3.6)$$

$$\theta^{12}(L) = 0, \quad \theta^{21}(L) = 0 \quad (3.7)$$

we shall obtain:

$$\phi^{11}(L) y_t + \phi^{12}(L) u_t = \theta^{11}(L) \varepsilon_t \quad (3.8)$$

$$\phi^{22}(L) u_t = \theta^{22}(L) a_t \quad (3.9)$$

The decomposition of (3.5) in (3.8) and (3.9) can also be obtained without imposing the restriction $\theta^{12}(L) = 0$. In fact, in this situation (3.5) becomes

$$\phi^{11}(L) y_t + \phi^{12}(L) u_t = \theta^{11}(L) \varepsilon_t + \theta^{12}(L) a_t \quad (3.10)$$

$$\phi^{22}(L) u_t = \theta^{22}(L) a_t \quad (3.11)$$

If we substitute the expression for a_t given by (3.11) into (3.10) and then rearrange terms, we obtain

$$\left| \theta^{22}(L) \right| \phi^{11}(L) y_t + \left[\left| \theta^{22}(L) \right| \phi^{12}(L) - \theta^{12}(L) \theta^{22}_{ad}(L) \phi^{22}(L) \right] u_t =$$

$$= \left| \theta^{22}(L) \right| \theta^{11}(L) \varepsilon_t \quad (3.12)$$

where $\theta^{22}_{ad}(L)$ is the adjoint matrix of $\theta^{22}(L)$ and $\left| \theta^{22}(L) \right|$ its determinant. We see then that equation (3.12) is of the same form as (3.8).

Therefore, the conditions:

$$\phi^{21}(L) = 0, \quad \theta^{21}(L) = 0$$

are, in general, sufficient but not necessary to be able to affirm the absence of causality from y_t to u_t.

The necessary and sufficient condition for y_t to not cause u_t is, according to Kang (1981), given by

$$\phi^{21}(L)\theta^{11}(L) - \phi^{11}(L)\theta^{21}(L) = 0$$

We will then observe that both (3.8) and (3.12) coincide with the structural form of the econometric model given by (2.1), and (3.9) demonstrates that, in general, the vector for the r exogenous variables is generated by the multivariate ARMA model (3.11), which we shall write as:

$$\bar{\phi}(L)u_t = \bar{\theta}(L)a_t \qquad (3.13)$$

with:

$$\bar{\phi}(L) = I + \sum_{i=1}^{p_u} \bar{\phi}_i L^i, \quad \bar{\theta}(L) = \sum_{i=0}^{q_u} \bar{\theta}_i L^i \qquad (3.14)$$

and where a_t is a white noise process such that:

$$E[a_t] = 0, \quad E[a_{t_1} a'_{t_2}] = \Sigma_a \delta_{t_1 t_2} \qquad (3.15)$$

Note that the joint consideration of the structural form of the econometric model, given by (2.1), and the model of the exogenous variables, given by (3.13), completes the formulation, because it supplies the (1+r) equations necessary for determining the (1+r) endogenous and exogenous variables.

In a way similar to that employed in the previous chapter, we can rewrite (3.13) in state-space form as follows:

$$x^b_{t+1} = \Phi^b x^b_t + E^b w^b_t \qquad (3.16)$$

$$u_t = H^b x^b_t + C^b v^b_t \qquad (3.17)$$

In this case:

$$\Phi^b = \begin{bmatrix} -\bar{\phi}_1 & I & 0 & \cdots & 0 \\ -\bar{\phi}_2 & 0 & I & \cdots & 0 \\ \cdot & \cdot & \cdot & \cdots & \cdot \\ \cdot & \cdot & \cdot & \cdots & \cdot \\ \cdot & \cdot & \cdot & \cdots & \cdot \\ -\bar{\phi}_{h-1} & 0 & 0 & \cdots & I \\ -\bar{\phi}_h & 0 & 0 & \cdots & 0 \end{bmatrix}, \quad E^b = \begin{bmatrix} \bar{\theta}_1 - \bar{\phi}_1\bar{\theta}_0 \\ \bar{\theta}_2 - \bar{\phi}_2\bar{\theta}_0 \\ \cdot \\ \cdot \\ \cdot \\ \bar{\theta}_h - \bar{\phi}_h\bar{\theta}_0 \end{bmatrix} \quad (3.18)$$

where:

$$H^b = \begin{bmatrix} I & 0 & \cdots & 0 \end{bmatrix}, \quad C^b = \bar{\theta}_0 \quad (3.19)$$

and:

$$h = \max(p_u, q_u)$$

The corresponding white noise processes would be given by:

$$w_t^b = v_t^b = a_t \quad (3.20)$$

and their covariance matrices would be:

$$Q^b = R^b = S^b = \Sigma_a \quad (3.21)$$

Thus, equations (3.16) and (3.17), with definitions (3.18)-(3.21), give the state-space formulation for the vector of exogenous variables.

3.2. State-Space Formulation.

The basic idea for obtaining an adequate parametrization of the econometric model with measurement errors in the endogenous and exogenous variables is to form a state-space model in which the observable variables are precisely the endogenous and exogenous variables of the original model. This is an idea used by Mehra (1976) in the context of the regression model with measurement errors.

Taking the state-space formulation of the econometric model given in (2.14) and (2.15), and the corresponding one for the exogenous variables in (3.16) and (3.17), we can go on to combine these in the following form:

$$\begin{bmatrix} x^a_{t+1} \\ x^b_{t+1} \end{bmatrix} = \begin{bmatrix} \Phi^a & \Gamma^a H^b \\ 0 & \Phi^b \end{bmatrix} \begin{bmatrix} x^a_t \\ x^b_t \end{bmatrix} + \begin{bmatrix} \Gamma^a C^b v^b_t + E^a w^a_t \\ E^b w^b_t \end{bmatrix} \quad (3.22)$$

$$\begin{bmatrix} y_t \\ u_t \end{bmatrix} = \begin{bmatrix} H^a & D^a H^b \\ 0 & H^b \end{bmatrix} \begin{bmatrix} x^a_t \\ x^b_t \end{bmatrix} + \begin{bmatrix} D^a C^b v^b_t + C^a v^a_t \\ C^b v^b_t \end{bmatrix} \quad (3.23)$$

We are now in a position to take into account the possibility of measurement errors in the variables as in (3.1) and (3.2). To do this, we shall modify the observation equation (3.23) to take these errors into account:

$$\begin{bmatrix} y^*_t \\ u^*_t \end{bmatrix} = \begin{bmatrix} H^a & D^a H^b \\ 0 & H^b \end{bmatrix} \begin{bmatrix} x^a_t \\ x^b_t \end{bmatrix} + \begin{bmatrix} D^a C^b v^b_t + C^a v^a_t + v_{yt} \\ C^b v^b_t + v_{ut} \end{bmatrix} \quad (3.24)$$

Equations (3.22) and (3.24) correspond finally to the desired parametrization. These can be re-written in condensed state-space form as follows:

$$x_{t+1} = \Phi x_t + E w_t \quad (3.25)$$

$$z_t = H x_t + C v_t \quad (3.26)$$

where:

$$x_{t+1} = \begin{bmatrix} (x^a_{t+1})' & (x^b_{t+1})' \end{bmatrix}' \quad (3.27)$$

$$z_t = \begin{bmatrix} (y^*_t)' & (u^*_t)' \end{bmatrix}' \quad (3.28)$$

$$\Phi = \begin{bmatrix} \Phi^a & \Gamma^a H^b \\ 0 & \Phi^b \end{bmatrix}, \quad E = \begin{bmatrix} E^a & \Gamma^a C^b \\ 0 & E^b \end{bmatrix} \qquad (3.29)$$

$$H = \begin{bmatrix} H^a & D^a H^b \\ 0 & H^b \end{bmatrix}, \quad C = \begin{bmatrix} C^a & D^a C^b & I & 0 \\ 0 & D^b & 0 & I \end{bmatrix} \qquad (3.30)$$

The two noise processes are:

$$w_t = \begin{bmatrix} \varepsilon_t' & a_t' \end{bmatrix}' \qquad (3.31)$$

$$v_t = \begin{bmatrix} \varepsilon_t' & a_t' & v_{yt}' & v_{ut}' \end{bmatrix}' \qquad (3.32)$$

and they are characterized by:

$$E[w_t] = 0, \quad E[v_t] = 0 \qquad (3.33)$$

$$E\left\{ \begin{bmatrix} w_{t_1} \\ v_{t_1} \end{bmatrix} \begin{bmatrix} w_{t_2}' & v_{t_2}' \end{bmatrix} \right\} = \begin{bmatrix} Q & S \\ S' & R \end{bmatrix} \delta_{t_1 t_2} \qquad (3.34)$$

It is generally reasonable to assume that the processes v_{yt}, v_{ut} are mutually uncorrelated with one another and with ε_t and a_t, therefore:

$$Q = \begin{bmatrix} \Sigma_\varepsilon & 0 \\ 0 & \Sigma_a \end{bmatrix}, \quad S = \begin{bmatrix} \Sigma_\varepsilon & 0 & 0 & 0 \\ 0 & \Sigma_a & 0 & 0 \end{bmatrix} \qquad (3.35)$$

$$R = \begin{bmatrix} \Sigma_\varepsilon & 0 & 0 & 0 \\ 0 & \Sigma_a & 0 & 0 \\ 0 & 0 & \Sigma_y & 0 \\ 0 & 0 & 0 & \Sigma_u \end{bmatrix} \qquad (3.36)$$

The matrices Φ^a, Γ^a, E^a, H^a, D^a and C^a are defined in (2.26) and (2.27) as functions of the parameters of the econometric model, and the matrices Φ^b, E^b, H^b and C^b in (3.18) and (3.19) are defined as functions of the parameters of the multivariate ARMA model for the exogenous variables given by (3.13). To summarize, equations (3.25) to (3.36) define the model to be estimated, and in it all the parameters for the econometric model will appear explicitly, including the covariance matrices Σ_y and Σ_u of the measurement errors of the endogenous and exogenous variables.

The formulation of the econometric model given by (2.1) and the process which generates the exogenous variables given by (3.13) implies, as we have seen in Section 3.1, introducing certain restrictions in the multivariate ARMA model corresponding to the compound vector $[y_t' \; u_t']'$. In this work we shall not treat the empirical selection of these restrictions and we will take the specification for (2.1) and (3.13) as known. That is to say, we shall limit ourselves exclusively to the problem of estimation and verification of the econometric model with measurement errors. Nevertheless, it is important to point out that, in this context, the explicit consideration of the observation errors v_{yt}, v_{ut}, which may be ARMA in general, see Section 5.3, can avoid situations with spurious feedback between y_t and u_t due to the feedback between the observed variables y_t^* and u_t^* caused by observation errors, see Newbold (1978).

Note that the model given by (3.25)-(3.26) is a particular case of the general state-space formulation given by (2.7)-(2.8), where $\Gamma = 0$, $D = 0$. Nonetheless, in the next chapter we shall deal with the estimation of the general formulation. Thus we shall obtain an estimation algorithm for the more general case, applicable to other situations and, in particular, to the case when not all the exogenous variables are observed with error. It would not be convenient, in this last case, to include the exogenous variables observed without error in the z_t vector of the observation equation, because this would unnecessarily increase the dimension of this vector and, as we shall see, the estimation algorithm requires the inversion of matrices whose order is precisely that of the dimension of z_t.

CHAPTER 4

ESTIMATION OF ECONOMETRIC MODELS WITH MEASUREMENT ERRORS

Having formulated the econometric model with errors in the variables in state-space, according to the following:

$$x_{t+1} = \Phi x_t + \Gamma u_t + E w_t \tag{4.1}$$

$$z_t = H x_t + D u_t + C v_t \tag{4.2}$$

with

$$E\left[w_t\right] = 0, \quad E\left[v_t\right] = 0$$

$$E\left\{\begin{bmatrix} w_{t_1} \\ v_{t_1} \end{bmatrix} \begin{bmatrix} w'_{t_2} & v'_{t_2} \end{bmatrix}\right\} = \begin{bmatrix} Q & S \\ S' & R \end{bmatrix} \delta_{t_1 t_2}$$

we shall go on to estimate the vector θ, containing p parameters, in which we shall include all the unknown elements within the following matrices: Φ, Γ, E, H, D, C, Q, R and S. Recall that vectors x_t, u_t and z_t have dimensions n, r, and m, respectively.

In general, the system given above can be such that its matrices depend explicitly upon time, that is to say: $\Phi(t,\theta)$, $\Gamma(t,\theta)$, etc.

We shall represent the sampling information by $\{Z^N, U^N\}$ where:

$$Z^N = \{z'_1, \ldots, z'_N\}'$$

$$U^N = \{u'_1, \ldots, u'_N\}'$$

The maximum likelihood estimator $\hat{\theta}$ of θ will be given by:

$$\hat{\theta}_N = \arg\max_\theta L_N(\theta) = \arg\max_\theta p(Z^N; U^N, \theta) \tag{4.3}$$

where $p(Z^N;U^N,\theta)$ is the joint probability density function of all the observations Z^N. Considered as a function of θ, this is the likelihood function $L_N(\theta)$.

Two basic problems with this approach arise in the calculation of $\hat{\theta}_N$. The first is in obtaining an expression for $p(Z^N;U^N,\theta)$, and the second in maximizing this expression with respect to θ. We shall turn to these in what follows.

In Section 4.1 we formulate the likelihood function for the econometric model with measurement errors in the variables. The evaluation of this function requires the estimation of the state vector as well as the calculation of the innovation process and its covariance matrix. This is achieved using the Kalman filter and through the solution of a matrix Riccati equation. In the situation, found frequently in econometrics, with constant coefficients and with a state vector larger than the vector of observed variables, it is convenient to substitute the Riccati equation for equations of the Chandrasekhar type, see Appendix A.

Section 4.2 turns to the question of obtaining the analytical expressions for the gradient and hessian of the likelihood function. The exact value of the information matrix is obtained from these. Next, Section 4.3 puts forward several procedures for dealing with the problem of initial conditions we have to use in the recursive scheme given by the Kalman filter.

The specific optimization algorithm used in the maximization of the likelihood function is described in Section 4.4 . In this context, the problem of identification of the model, by means of the analysis of the eigenvalues and eigenvectors of the information matrix, is also examined.

In Section 4.5 we review the recent literature on asymptotic properties of the maximum likelihood estimator for the case of dependent observations.

Section 4.6 includes some numerical considerations with regard to the use of adimensional expressions for the gradient and hessian, and on the Cholesky factorization of the covariance matrices to assure the positive semidefinite property.

Finally, we shall examine aspects of the verification of the model in Section 4.7. The fact that the Lagrange-multiplier test can be constructed with expressions that have already been calculated in the process of estimation is emphasized.

4.1. Evaluation of the Likelihood Function.

The basic difficulty in the calculation of the likelihood function is that the observations are not independent, that is:

$$p(Z^N;U^N,\theta) \neq p(z_1;U^1,\theta) \ldots p(z_N;U^N,\theta)$$

The central idea for solving this problem, see Schweppe (1965) and Harvey (1981a, 1981b), is to apply the definition of conditional densities successively in writing the expression for $p(Z^N;U^N,\theta)$, so that

$$p(Z^N;U^N,\theta) = p(z_N|Z^{N-1};U^{N-1},\theta)p(z_{N-1}|Z^{N-2};U^{N-2},\theta) \ldots p(z_1;U^1,\theta) \quad (4.4)$$

where $p(z_t|Z^{t-1};U^{t-1},\theta)$ is the conditional probability density of the current observations, z_t, given past observations Z^{t-1}.

The factorization thus obtained can be calculated given that, under the assumption of normality in the perturbations w_t and v_t, the general term $p(z_t|Z^{t-1};U^{t-1},\theta)$ is characterized completely by means of the Kalman filter applied to the system for any given value of θ. In effect, the following notation:

$$E\left[z_t|Z^{t-1};U^{t-1},\theta\right] = \hat{z}_{t|t-1}$$

$$\text{cov}\left[z_t|Z^{t-1};U^{t-1},\theta\right] = E\left[(z_t - \hat{z}_{t|t-1})(z_t - \hat{z}_{t|t-1})'\right] = B_t \quad (4.5)$$

renders:

$$p(z_t|Z^{t-1};U^{t-1},\theta) = (2\pi)^{-m/2} |B_t|^{-1/2} e^{-1/2\, \tilde{z}_t' B_t^{-1} \tilde{z}_t} \quad (4.6)$$

where $\tilde{z}_t$ is termed the innovation process, defined as:

$$\tilde{z}_t = z_t - \hat{z}_{t|t-1} \tag{4.7}$$

The expressions for $\tilde{z}_t$ and B_t are obtained from the following Kalman filter, see Appendix A:

$$\tilde{z}_t = z_t - H\hat{x}_{t|t-1} - Du_t \tag{4.8}$$

$$\hat{x}_{t+1|t} = \Phi\hat{x}_{t|t-1} + \Gamma u_t + K_t\tilde{z}_t \tag{4.9}$$

$$K_t = \left[\Phi P_{t|t-1}H' + ESC'\right] B_t^{-1} \tag{4.10}$$

$$P_{t+1|t} = \Phi P_{t|t-1}\Phi' + EQE' - K_t B_t K_t' \tag{4.11}$$

$$B_t = HP_{t|t-1}H' + CRC' \tag{4.12}$$

From the factorization given by (4.4) and the notation in (4.3), we can write:

$$L_N(\theta) = p(Z^N; U^N, \theta) = \prod_{t=1}^{N} p(z_t | Z^{t-1}; U^{t-1}, \theta) \tag{4.13}$$

We now have to maximize the previous expression with respect to the vector θ. This is equivalent to minimizing the negative log-likelihood function:

$$\ell_N(\theta) = -\log p(Z^N; U^N, \theta) = -\sum_{t=1}^{N} \log p(z_t | Z^{t-1}; U^{t-1}, \theta)$$

$$= \sum_{t=1}^{N} \left[\frac{m}{2}\log(2\pi) + \frac{1}{2}\log|B_t| + \frac{1}{2}\tilde{z}_t' B_t^{-1}\tilde{z}_t\right] \tag{4.14}$$

Note that by ignoring the constant, the expression $\ell(\theta)$ consists of two terms: a deterministic term which depends on B_t and a quadratic term in $\tilde{z}_t$, which recalls the principle of least squares.

From now on, and in order to simplify the notation, we shall eliminate subindex N. In this way, for example, $\ell_N(\theta)$ or $\hat{\theta}_N$ will be written as $\ell(\theta)$ and $\hat{\theta}$, respectively.

Thus the problem is reduced to minimizing expression (4.14), subject to the restrictions introduced by the Kalman filter and reflected in equations (4.8) to (4.12).

It is formally a mathematical program with restrictions, although soluble by substitution in the objective function. That is the reason why the procedures described in Section 4.4 can be applied to it. We are looking for a solution $\hat{\theta}$ of the equation $\frac{\partial \ell(\theta)}{\partial \theta} = 0$ with $\frac{\partial^2 \ell(\theta)}{\partial \theta \partial \theta'} > 0$.

Note that the problem of estimating the parameters of the system (4.1)-(4.2) requires the estimation of the state vector given by the Kalman filter. In our case, and given that we do not know the true value of θ but only have an estimate, the estimates derived from the Kalman filter and used in successive evaluations of the likelihood function are suboptimal. Nonetheless, if the process of estimation is such that $\hat{\theta} \to \theta$, then the value of $z_t - \hat{z}_{t|t-1}$ would tend to the true innovation process.

We should recall that the results obtained up to this point are also valid for time-varying systems.

In the case of systems with constant coefficients, the expressions (4.10), (4.11), and (4.12) for the Kalman filter can be given alternative expressions which, in some situations, are more efficient from a computational point of view. In fact, the solution of the Riccati matrix equation (4.11) amounts to the propagation of $[n(n + 1)]/2$ equations. Unfortunately it is common in the practice of econometrics to find large values for n, and this is certainly true for seasonal time series models.

Thus, for example, the seasonal autoregressive process:

$$(1 - \bar{\phi}_1 B^{12} - \bar{\phi}_2 B^{24}) y_t = \varepsilon_t$$

corresponds to a state-space model with n = 24. Examples of this type are characterized by large values of n, and much larger than the dimension m of the vector of observable variables, in this case m = 1.

Among computationally efficient procedures for models with constant coefficients and n >> m are those which replace the nonlinear Riccati equation given by (4.11) by certain nonlinear difference equations of the Chandrasekhar type, see Anderson and Moore (1979). With these formulations, the number of equations may vary with 2mn, rather than with $n^2/2$.

Appendix A analyses the Chandrasekhar formulations, and situations where these formulations offer computational advantages. Their use, with the initial conditions we shall refer to later on, is advisable provided that the model to be estimated, as given by (4.1) and (4.2), is such that n >> m.

Chan et al. (1984) have recently extended the classical results on the stability of the Kalman filter given by (4.8)-(4.12) to nonstationary models with unit roots. Thus the estimation procedure proposed here can be applied to models with unit roots. Therefore the order of differencing to transform variables for stationarity need not be fixed *a priori*.

The likelihood function for a nonstationary process is not defined in the usual form because the density only exists conditionally. It is not possible to define the likelihood function uniquely for a non-stationary model without first specifying initial conditions. We will later consider the determination of initial conditions for the Kalman filter in nonstationary cases.

4.2. **Maximization of the Likelihood Function.**

The minimization of expression (4.14) requires numerical optimization procedures. Note that the restrictions on this problem given by the Kalman filter do not apply to the vector of parameters θ to be estimated, but rather to expressions which are implicit functions of θ.

There are a number of different optimization methods applicable in this case. All of them use first derivative information on $\ell(\theta)$, and some of them require second derivative information. Therefore, the problem that arises now is the calculation of $\frac{\partial \ell(\theta)}{\partial \theta}$ and $\frac{\partial^2 \ell(\theta)}{\partial \theta \partial \theta'}$.

All of the results relating to $\frac{\partial \ell(\theta)}{\partial \theta}$, which we shall detail shortly, and which are not immediately apparent, are set out in Appendix B.

If θ_i is the i-th element of the vector θ, taking the partial derivative in (4.14) with respect to θ_i, we see that:

$$\frac{\partial \ell(\theta)}{\partial \theta_i} = \sum_{t=1}^{N} \left\{ -\frac{1}{2} \operatorname{tr}\left[B_t^{-1} \frac{\partial B_t}{\partial \theta_i}\right] + \tilde{z}_t' B_t^{-1} \frac{\partial \tilde{z}_t}{\partial \theta_i} - \frac{1}{2} \tilde{z}_t' B_t^{-1} \frac{\partial B_t}{\partial \theta_i} B_t^{-1} \tilde{z}_t \right\}$$

(4.16)

We now have to calculate $\dfrac{\partial \tilde{z}_t}{\partial \theta_i}$ and $\dfrac{\partial B_t}{\partial \theta_i}$. From (4.8) we get:

$$\frac{\partial \tilde{z}_t}{\partial \theta_i} = -\frac{\partial H}{\partial \theta_i} \hat{x}_{t|t-1} - H \frac{\partial \hat{x}_{t|t-1}}{\partial \theta_i} - \frac{\partial D}{\partial \theta_i} u_t \qquad (4.17)$$

The sensitivity equations of the filter are obtained by differentiating (4.9), so that:

$$\frac{\partial \hat{x}_{t+1|t}}{\partial \theta_i} = \frac{\partial \bar{\Phi}_t}{\partial \theta_i} \hat{x}_{t|t-1} + \bar{\Phi}_t \frac{\partial \hat{x}_{t|t-1}}{\partial \theta_i} + \frac{\partial \bar{\Gamma}_t}{\partial \theta_i} u_t + \frac{\partial K_t}{\partial \theta_i} z_t \qquad (4.18)$$

where:

$$\bar{\Phi}_t = \Phi - K_t H, \qquad \bar{\Gamma}_t = \Gamma - K_t D \qquad (4.19)$$

From (4.10) we can also deduce that:

$$\frac{\partial K_t}{\partial \theta_i} = \left[\frac{\partial \Phi}{\partial \theta_i} P_{t|t-1} H' + \Phi \frac{\partial P_{t|t-1}}{\partial \theta_i} H' + \Phi P_{t|t-1} \frac{\partial H'}{\partial \theta_i} \right.$$

$$+ \frac{\partial E}{\partial \theta_i} SC' + E \frac{\partial S}{\partial \theta_i} C' + ES \frac{\partial C'}{\partial \theta_i} \bigg] B_t^{-1}$$

$$- \left[\Phi P_{t|t-1} H' + ESC' \right] B_t^{-1} \frac{\partial B_t}{\partial \theta_i} B_t^{-1} \qquad (4.20)$$

Differentiating (4.12) we get:

$$\frac{\partial B_t}{\partial \theta_i} = \frac{\partial H}{\partial \theta_i} P_{t|t-1} H' + H \frac{\partial P_{t|t-1}}{\partial \theta_i} H' + H P_{t|t-1} \frac{\partial H'}{\partial \theta_i}$$

$$+ \frac{\partial C}{\partial \theta_i} RC' + C \frac{\partial R}{\partial \theta_i} C' + CR \frac{\partial C'}{\partial \theta_i} \qquad (4.21)$$

And, after some algebraic manipulation, we can obtain from (4.11) the following expression for the sensitivity equations of the covariance matrix of the Kalman filter:

$$\frac{\partial P_{t+1|t}}{\partial \theta_i} = \bar{\Phi}_t \frac{\partial P_{t|t-1}}{\partial \theta_i} \bar{\Phi}'_t + A_{it} + A'_{it} \qquad (4.22)$$

where:

$$A_{it} = \frac{\partial \Phi}{\partial \theta_i} P_{t|t-1} \bar{\Phi}'_t - K_t \frac{\partial H}{\partial \theta_i} P_{t|t-1} \bar{\Phi}'_t$$

$$- \frac{\partial E}{\partial \theta_i} SC'K'_t - E \frac{\partial S}{\partial \theta_i} C'K'_t - ES \frac{\partial C'}{\partial \theta_i} K'_t$$

$$+ \frac{1}{2} \left[\frac{\partial E}{\partial \theta_i} QE' + E \frac{\partial Q}{\partial \theta_i} E' + EQ \frac{\partial E'}{\partial \theta_i} \right]$$

$$+ \frac{1}{2} K_t \left[\frac{\partial C}{\partial \theta_i} RC' + C \frac{\partial R}{\partial \theta_i} C' + CR \frac{\partial C'}{\partial \theta_i} \right] K'_t \qquad (4.23)$$

Thus we can see that the calculation of the gradient of the likelihood function requires the propagation of equations (4.9), (4.11), (4.18) and (4.22). If the vector of parameters θ is p-dimensional the computational burden for each of these expressions will be n, $[n(n+1)]/2$, np and $[n(n+1)]p/2$ equations, respectively.

There are several ways of reducing this computational burden. Normally this is done at the expense of an increase in storage requirements. It is not our aim to analyze problems of this nature in this work. But we would point out that an analysis of this type of problem would have to include details of the structure of the system's matrices as well as algorithms that used, explicitly, the specific form of the function to be minimized given by (4.14).

For the reasons given in Appendix A, it is advisable to use the following expression, instead of (4.11), in the numerical calculation of $\ell(\theta)$:

$$P_{t+1|t} = \bar{\Phi}_t P_{t|t-1} \bar{\Phi}_t' + \begin{bmatrix} E & -K_t C \end{bmatrix} \begin{bmatrix} Q & S \\ S' & R \end{bmatrix} \begin{bmatrix} E' \\ -C'K_t' \end{bmatrix}$$

In addition, it is necessary to point out that, in cases where the system's matrices do not depend on time, the calculations for obtaining $\frac{\partial \ell(\theta)}{\partial \theta}$ would be considerably simplified. In a case such as this, and under certain conditions, see Chan et al. (1984), it can be shown that the Kalman-filter matrices converge to their steady-state values given by P, K and B. This requires the solution to the steady-state version of (4.11) which we shall express as:

$$P = \Phi P \Phi' + EQE' - \begin{bmatrix} \Phi PH' + ESC' \end{bmatrix} \begin{bmatrix} HPH' + CRC' \end{bmatrix}^{-1} \begin{bmatrix} \Phi PH' + ESC' \end{bmatrix}'$$

(4.24)

This expression is a Riccati algebraic matrix equation.

Similarly, the steady-state expression corresponding to (4.22) would be:

$$\frac{\partial P}{\partial \theta_i} = \bar{\Phi} \frac{\partial P}{\partial \theta_i} \bar{\Phi}' + A_i + A_i' \qquad (4.25)$$

which is a Lyapunov algebraic equation.

Equations (4.24) and (4.25) can be solved efficiently by means of a series of procedures developed in the control theory literature, see Appendix F. The use of steady-state values is particularly justified if the optimal time varying Kalman filter reaches steady-state values in a time that is short relative to the time interval of the observations N.

As to the evaluation of the hessian of the likelihood function, it is necessary to differentiate expression (4.16). The expression corresponding to the general term $\frac{\partial^2 \ell(\theta)}{\partial \theta_i \partial \theta_j}$ is obtained in Appendix C, along with the computational requirements associated with its calculation.

An alternative approach to the calculation of the hessian is to use the following information matrix instead:

$$M(\theta) = E\left[\frac{\partial^2 \ell(\theta)}{\partial \theta \partial \theta'}\bigg|_\theta\right] = E\left\{\left[\frac{\partial \ell(\theta)}{\partial \theta}\right]\left[\frac{\partial \ell(\theta)}{\partial \theta}\right]'\bigg|_\theta\right\} \qquad (4.26)$$

or, as is frequently done, an approximation thereof.

The information matrix is such that its inverse gives an asymptotic lower bound to the covariance matrix of the parameter estimates.

In Appendix D we obtain the following expression for the general term of the information matrix:

$$[M(\theta)]_{ij} = \sum_{t=1}^{N} \left\{ \frac{1}{2} \mathrm{tr}\left[B_t^{-1} \frac{\partial B_t}{\partial \theta_i} B_t^{-1} \frac{\partial B_t}{\partial \theta_j}\right] + \mathrm{tr}\left[B_t^{-1} E\left[\frac{\partial \tilde{z}_t}{\partial \theta_i} \frac{\partial \tilde{z}_t'}{\partial \theta_j}\right]\right] \right\}$$

$$(4.27)$$

or:

$$[M(\theta)]_{ij} = \sum_{t=1}^{N} \left\{ \frac{1}{2} \mathrm{tr}\left[B_t^{-1} \frac{\partial B_t}{\partial \theta_i} B_t^{-1} \frac{\partial B_t}{\partial \theta_j}\right] + E\left[\frac{\partial \tilde{z}_t'}{\partial \theta_i} B_t^{-1} \frac{\partial \tilde{z}_t}{\partial \theta_j}\right] \right\}$$

$$(4.28)$$

This is the expression obtained by Engle and Watson (1981). Watson and Engle (1983) use the following approximation:

$$[M(\theta)]_{ij} = \sum_{t=1}^{N} \left\{ \frac{1}{2} \text{tr}\left[B_t^{-1} \frac{\partial B_t}{\partial \theta_i} B_t^{-1} \frac{\partial B_t}{\partial \theta_j} \right] + \frac{\partial \tilde{z}_t'}{\partial \theta_i} B_t^{-1} \frac{\partial \tilde{z}_t}{\partial \theta_j} \right\} \quad (4.29)$$

Note that this approximation amounts to using a particular realization of the random term on the right-hand side of (4.28) instead of its expected value. Thus it is valid only to the extent that its standard deviation is much smaller than its expected value. In any case, this will depend on the specific characteristics of the model to be estimated.

Because, as we have pointed out, the information matrix plays a fundamental role in our analysis, it is important to work with its exact expression. Given that:

$$E\left[\frac{\partial \tilde{z}_t}{\partial \theta_i} \frac{\partial \tilde{z}_t'}{\partial \theta_j} \right] = B_t^{ij} + \frac{\overline{\partial \tilde{z}_t}}{\partial \theta_i} \frac{\overline{\partial \tilde{z}_t'}}{\partial \theta_j} \quad (4.30)$$

where $\dfrac{\overline{\partial \tilde{z}_t}}{\partial \theta_i}$ is the mean of $\dfrac{\partial \tilde{z}_t}{\partial \theta_i}$ and B_t^{ij} is the covariance matrix of $\dfrac{\partial \tilde{z}_t}{\partial \theta_i}$ and $\dfrac{\partial \tilde{z}_t}{\partial \theta_j}$, we can obtain from (4.27) and (4.30) the following expression for the information matrix:

$$[M(\theta)]_{ij} = \sum_{t=1}^{N} \left\{ \frac{1}{2} \text{tr}\left[B_t^{-1} \frac{\partial B_t}{\partial \theta_i} B_t^{-1} \frac{\partial B_t}{\partial \theta_j} \right] + \text{tr}\left[B_t^{-1} B_t^{ij} \right] \right.$$

$$\left. + \text{tr}\left[B_t^{-1} \frac{\overline{\partial \tilde{z}_t}}{\partial \theta_i} \frac{\overline{\partial \tilde{z}_t'}}{\partial \theta_j} \right] \right\} \quad (4.31)$$

The evaluation of the vectors $\dfrac{\partial \tilde{z}_t}{\partial \theta_i}$ and $\dfrac{\partial \tilde{z}_t}{\partial \theta_j}$ and the matrix B_t^{ij} requires the propagation of the equations for the mean and covariance matrix of a linear system having dimension 3n. The formulation of this system and of the corresponding propagation equations is developed in Appendix D.

Note that both the exact expression of the information matrix given by (4.31) and its approximation given by (4.29) only use the first derivatives of the innovation process and of its covariance matrix.

If instead of using the matrix equations (4.10), (4.11) and (4.12), for the calculation of K_t, $P_{t+1|t}$ and B_t we use algorithms of the Chandrasekhar type, as suggested in Appendix A and given by expressions (A.21) to (A.24), we should have to alter some of the expressions we obtained for the calculation of the gradient. This would be done by substituting expressions (4.20), (4.21) and (4.22) by those obtained by deriving expressions (A.21), (A.22), (4.23) and (A.24) with respect to θ_i. Otherwise, and especially in the case of the calculation of the hessian and information matrix the previous results are also valid.

At this point we have obtained all the analytical expressions needed to evaluate the likelihood function, its gradient, its hessian and the corresponding information matrix.

4.3. Initial Conditions.

To begin the iterative process implicit in this set of equations we need an initial value for the parameters to be estimated θ^0, and for the initial conditions of the system set out in (4.1)-(4.2) which would allow the algorithm of the Kalman filter given by equations (4.8)-(4.12) to be started. We shall refer to the problem of the choice of θ^0 in Section 4.6.

If the initial state of the system is x_0, we will represent these initial conditions by $\bar{x}_0$ and P_0, so that:

$$\bar{x}_0 = E[x_0], \quad P_0 = E\left\{[x_0 - \bar{x}_0][x_0 - \bar{x}_0]'\right\}$$

There are several approaches to determining $\bar{x}_0$ and P_0. See, for example, Harvey and Todd (1983), Ansley and Kohn (1983), and Rosenberg (1973).

The simplest and most direct is to set $\bar{x}_0 = 0$ and $P_0 = \tau I$ where $\tau \gg 0$. Using equations (4.8) and (4.12) of the Kalman filter it is possible to obtain, with these initial conditions, the values for $\tilde{z}_n$ and B_n, and the likelihood function which we shall maximize will correspond to the minimization of:

$$\ell_{N-n}(\theta) = \sum_{t=n+1}^{N} \left[\frac{m}{2}\log(2\pi) + \frac{1}{2}\log|B_t| + \frac{1}{2}\tilde{z}_t' B_t^{-1} \tilde{z}_t \right]$$

This procedure can be used in stationary as well as nonstationary models. The first n observations are used for obtaining the initial conditions for the Kalman filter. This method presents in practice numerical problems which can be solved using an alternative formulation of the Kalman filter which propagates the inverse matrix of $P_{t+1|t}$ known as the information filter, see Anderson and Moore (1979, pp. 138-142). However, in many econometric models the transition matrix Φ may be singular and thus may not have the inverse required by the information filter.

For reasons given previously, we shall use another two procedures for the determination of the initial conditions.

When the system is stationary, the initial conditions are given by:

$$\bar{x}_0 = [I - \Phi]^{-1} \Gamma \bar{u}_0 \qquad (4.32)$$

In general, when there are no constant terms in the model $\bar{u}_0 = 0$, then $\bar{x}_0 = 0$. The P_0 matrix will be given by the solution of the algebraic Lyapunov equation:

$$P = \Phi P \Phi' + EQE' \qquad (4.33)$$

That is to say $P_0 = P$. As we have pointed out, there are numerically efficient procedures for the calculation of P, see Appendix F.

Where nonstationary systems are concerned, we cannot refer to the previous state of equilibrium, but rather we can consider x_0 as an unknown parameter. Under this assumption, $P_0 = 0$, and we have to obtain an estimate of x_0. Appendix E develops this algorithm, and analyses the conditions under which the initial state x_0 is identifiable.

It is appropriate to point out that the use of these initial conditions is especially suited to initiating not only the Kalman filter as given by (4.8) to (4.12), but also to initiating the Chandrasekhar equations, as is shown in Appendix A.

4.4. Gradient Methods and Identification.

Once the expressions for the gradient and hessian of the function to be minimized are obtained, we are in a position to refer to the specific optimization methods employed. We shall basically consider gradient methods and some of their variations that tend to avoid numerical problems.

As is known, the basic form of the gradient methods is as follows:

$$\theta^{i+1} = \theta^i - \rho^i w^i \left.\frac{\partial \ell(\theta)}{\partial \theta}\right|_{\theta=\theta^i}$$

where θ^i is the value for the vector of the parameters to be estimated at the i-th iteration, w^i is an approximation to the inverse of the hessian, $[J(\theta^i)]^{-1}$, and ρ^i is a scalar which is determined by a one-dimensional search procedure with the aim of ensuring that $\ell(\theta^{i+1}) < \ell(\theta^i)$. It can be shown, see Luenberger (1984), that convergence near the minimum is characterized by:

$$\ell(\theta^{i+1}) < \left[\frac{\lambda_{max} - \lambda_{min}}{\lambda_{max} + \lambda_{min}}\right]^2 \ell(\theta^i)$$

where λmax and λmin are the maximum and minimum eigenvalues of W, respectively.

The different optimization methods employed correspond to different choices of W^i and, in some cases, of ρ^i and $\frac{\partial \ell(\theta)}{\partial \theta}$. See Luenberger (1984), Gill et al. (1981), and Quandt (1983).

The Newton-Raphson method, in particular, corresponds to the choice:

$$W^i = \left[\frac{\partial^2 \ell(\theta)}{\partial \theta \, \partial \theta'} \bigg|_{\theta=\theta^i} \right]^{-1} = \left[J(\theta^i) \right]^{-1} \tag{4.34}$$

This is undoubtfully the procedure that has the best convergence properties, but it involves three basic difficulties. It will not converge when the hessian ceases to be a positive definite matrix, and numerical problems arise when the hessian is nearly singular. In addition, its calculation, as shown in Appendix C, has very high computational requirements.

The method we shall employ is basically one known in the literature as the method of "scoring", which corresponds to the following choice:

$$W^i = \left\{ E\left[\frac{\partial^2 \ell(\theta)}{\partial \theta \, \partial \theta'} \right] \bigg|_{\theta=\theta^i} \right\}^{-1} = \left[M(\theta^i) \right]^{-1} \tag{4.35}$$

In this case the W^i matrix is the inverse of the information matrix. The exact expression for this matrix is given by (4.31) and an approximation is given by (4.29). In both cases, only the first derivatives of the innovation process and of its covariance matrix are used, which substantially reduces the computational requirements for the calculation of the hessian. Given that the matrix $M(\theta^i)$ is semidefinite positive, a value for ρ^i can always be obtained so that $\ell(\theta^{i+1}) < \ell(\theta^i)$.

Although this procedure avoids problems involved in the calculation of the hessian, difficulties arise when the information matrix is singular, and this matrix can even become negative definite in such a way that $\ell(\theta^{i+1}) > \ell(\theta^i)$ for any value of ρ^i. Moreover, in this situation, the step size given by $\Delta\theta^i = -\rho^i \left[M(\theta^i)\right]^{-1} \left.\frac{\partial \ell(\theta)}{\partial \theta}\right|_{\theta=\theta^i}$ is very large in the direction of the eigenvectors corresponding to the eigenvalues close to zero. In effect, if λ_i and v_i, $i = 1,\ldots,p$ are, respectively, the eigenvalues and eigenvectors of the information matrix M, we can write the following:

$$M(\theta^i) = \sum_{j=1}^{p} \lambda_j v_j v_j'$$

and thus:

$$\left[M(\theta^i)\right]^{-1} = \sum_{j=1}^{p} \frac{1}{\lambda_j} v_j v_j' \qquad (4.36)$$

that is:

$$\Delta\theta^i = -\rho^i \sum_{j=1}^{p} \frac{1}{\lambda_j} v_j' \left[\left.\frac{\partial \ell(\theta)}{\partial \theta}\right|_{\theta=\theta^i}\right] v_j$$

Thus we see that the step size in the direction of the eigenvector v_j corresponding to the eigenvalue λ_j would be very large for values of this eigenvalue close to zero. In other words, the algorithm takes large steps in those parameter directions about which poor information is available. Clearly this raises problems of convergence.

The procedure we have used, based on Gupta and Mehra (1974), to avoid the difficulties in the convergence of the algorithm in a situation where the information matrix is singular or nearly singular, is as follows. If we place the eigenvalues in the following order:

$$\lambda_1 > \lambda_2 > \ldots > \lambda_{p-k} > \ldots > \lambda_p$$

and we assume that λ_1/λ_{p-k} is the greatest admissible dispersion enabling the matrix $M(\theta^i)$ to be well-conditioned, we can make:

$$M(\theta^i) \simeq \sum_{j=1}^{p-k} \lambda_j v_j v_j'$$

and thus:

$$\left[M(\theta^i)\right]^{-1} \approx \sum_{j=1}^{p-k} \frac{1}{\lambda_j} v_j v_j' \qquad (4.37)$$

That is the inverse of the information matrix is computed leaving out one or more of the smallest eigenvalues.

The eigenvalues of the information matrix are the dimensions of the uncertainty hiperellipsoid of constant likelihood, associated with the parameter estimates, within the parameter space. The eigenvalues provide information on the volume of these surfaces and the smaller ones indicate a larger dimension and therefore more uncertainty. The eigenvectors give the directions of the axes of the hiperellipsoid.

The eigenvectors not included in (4.37) indicate the k linear combinations of the parameters of the vector θ which are not identifiable. Note that by doing this the linear combination of the non-identifiable parameters is precisely fixed, and that this differs from procedures used normally in econometrics, which place exclusion restrictions on the specification of the model or impose other arbitrary *a priori* restrictions on the parameters so as to make the model identifiable.

The basic method for updating the parameter vector will be:

$$\theta^{i+1} = \theta^i - \rho^i \sum_{j=1}^{p-k} \frac{1}{\lambda_j} v_j' \left[\frac{\partial \ell(\theta)}{\partial \theta}\bigg|_{\theta=\theta^i}\right] v_j \qquad (4.38)$$

Note that this reduces the dimension of the space in which we search for the parameter vector which minimizes $\ell(\theta)$, to that of the subspace corresponding to the (p - k) dominant eigenvalues. It should be pointed out, however, that by means of (4.38) all the parameters to be estimated are updated.

Once an *a priori* upper bound to the maximum admissible dispersion for the eigenvalues has been fixed, the program determines the increase in the vector θ in the (i + 1)-th iteration according to (4.38). By adding the eigenvalue λ_{p-k-1}, a new updating of θ can be obtained. The process is repeated to include successive eigenvalues up to λ_p, which implies the inversion of $M(\theta^i)$ according to (4.36). In this way, a set of (k + 1) values for the parameter vector θ^{i+1} is obtained for each iteration, and from them the one providing the minimum value of $\ell(\theta^{i+1})$ must be selected. We can then proceed to the next iteration.

Because the parameter vector θ is locally identifiable if, and only if, the information matrix $M(\theta)$ is not singular, see Bowden (1973) and Rothenberg (1971), the method used implies a test for the identification of the model. Moreover, $\left[M(\hat{\theta})\right]^{-1}$ represents a lower bound for the covariance matrix of the estimated vector $\hat{\theta}$.

However, it is important to point out that the eigenvalues of $M(\theta)$ can be quite sensitive to the sample size and to the particular realization used. For this reason we consider of special interest the method used for updating the parameter vector given by (4.38).

In this context, our opinion is that the criteria employed in the identification of dynamic econometric models with measurement errors that have appeared in recent years in the econometrics literature, see Aigner et al. (1984, pp. 1372-1386), are not of great use in practice. First, the results cannot be applied easily in general terms, for example, to simultaneous equations or to correlated observation errors. Second, they cannot be used to obtain efficient parameter estimates. Finally, when a model is not identifiable, it is not possible to obtain reasonable criteria, from an empirical point of view, which would allow restrictions to be imposed to make the model identifiable. We illustrate these points in Chapter 6.

4.5. Asymptotic Properties.

In this section we shall discuss the asymptotic properties (consistency, efficiency and asymptotic normality) of the maximum likelihood estimator.

The maximum likelihood approach was first proposed by Fisher (1921). Cramer (1946) proved the consistency of this estimator in the case of independent and identically distributed observations. To prove it, he assumes that the log-likelihood function is three times differentiable. This hypothesis of differentiability was avoided in Wald (1949).

The problem treated here does not correspond to the classical formulation, since the observations $Z^N = \{z_1', \ldots, z_N'\}'$ are neither identically nor independently distributed.

The different analyses of the consistency property of the maximum likelihood estimators for dependent observations have been carried out following either the work by Cramer (1946) or the one by Wald (1949). Among the first ones we should point out the works by Bar-Shalom (1971), Basawa et al. (1976) and above all the one by Crowder (1976). We could include within Wald's (1949) tradition the studies by Silvey (1961), Bhat (1974), Caines and Rissanen (1974), and Heijmans and Magnus (1986a, and 1986b).

The fundamental differences distinguishing these studies are due to the use of more or less restrictive assumptions, and to the different degree of difficulty in verifying their fulfilment.

In the context of linear state-space models, Pagan (1980) showed the consistency and asymptotic normality of $\hat{\theta}_N$, applying Crowder's (1976) conditions for the form given by (3.25)-(3.26) with z_t scalar. The generalization of this result to the situation where z_t is a vector is immediate, as Pagan (1980, p. 342) indicated. The hypotheses required to assure these properties are certain regularity conditions on $p(Z^n, \theta^*)$, the identifiability of the model and some conditions which characterize the convergence properties of $J = \dfrac{\partial^2 \ell(\theta)}{\partial \theta \partial \theta'}$ at θ^* and its neighbourhoods, where θ^* is the "true" parameter value.

We may summarize the asymptotic properties of the estimator $\hat{\theta}_N$ as follows.

Consistency:

$$\lim_{N \to \infty} \hat{\theta}_N = \theta^*$$

Asymptotic unbiasedness:

$$\lim_{N \to \infty} E\left[\hat{\theta}_N\right] = \theta^*$$

Asymptotic normality:

As $N \to \infty$, $\hat{\theta}_N$ tends to a gaussian random vector with mean θ^* and covariance matrix $[M(\theta^*)]^{-1}$.

Asymptotic efficiency:

$$\lim_{N \to \infty} \text{Cov}[\hat{\theta}_N] = [M(\theta^*)]^{-1}$$

Since the estimator is asymptotically unbiased the Cramer-Rao lower bound is reached.

The generalization of these results obtained for the model (3.25)-(3.26), to the general formulation given by (4.1)-(4.2) is analytically more complex, and also requires some additional assumptions on the input u_t. Caines (1988, Chapter 7) analyzes the asymptotic properties of $\hat{\theta}_N$ under the assumptions of deterministic and stochastic inputs. Under quite general conditions he shows the consistency and asymptotic normality of $\hat{\theta}_N$. Hannan and Deistler (1988, Chapter 4) avoid the compactness hypothesis in their analysis.

A natural generalization of the maximum likelihood estimator is the minimum prediction identification error method which does not use the hypotheses of linearity, normality and exact modelling in obtaining the appropriate generalized version of these asymptotic results, see Caines (1988, Chapter 8) and Ljung (1987, Chapter 9).

4.6. Numerical Considerations.

It is worth emphasizing some important considerations with respect to the numerical aspects of the minimization of $\ell(\theta)$.

Given that the parameters of the vector θ can have very different values, it is important to scale the parameters previously to reduce them to similar values. This amounts to working with adimensional expressions for the gradient and for the hessian given by the following general terms:

$$g_i^*(\theta) = \frac{1}{|\theta_i|} \frac{\partial \ell(\theta)}{\partial \theta_i}, \qquad J_{ij}^* = \frac{1}{|\theta_i||\theta_j|} \frac{\partial^2 \ell(\theta)}{\partial \theta_i \partial \theta_j}$$

Another consideration relates to the necessity of insuring, during the iterative minimization process, that the positive semidefinite and positive definite nature of the covariance matrices Q and R of the noise processes are maintained. We have achieved this by using the Cholesky factorization

$$Q = Q^*\left[Q^*\right]', \qquad R = R^*\left[R^*\right]'$$

where Q^* and R^* are lower triangular matrices. The program estimates the parameters of Q and R by introducing the conditions that $Q_{ii}^* \geq 0$ and $R_{ii}^* > 0$.

The choice of the initial value θ^0 matters in two ways. If it is too far away from a solution, the iterative procedure can diverge or converge to a merely local minimum. In general the algorithm defined by (4.38) guarantees the convergence to a local minimum of $\ell(\theta)$. Nevertheless, the equation:

$$\frac{\partial \ell(\theta)}{\partial \theta} = 0$$

may have multiple solutions and the value in which we are interested should correspond to a global minimum.

If $\ell(\theta)$ is a convex function, for all θ within the entire parameter space, the local minimum condition is guaranteed to yield a global minimum. There are few interesting cases for which this property can be proved analytically, and even if $\ell(\theta)$ has only one local minimum $\ell_N(\theta)$ may have other local minima due to the properties of the finite sample used.

To obtain the global minimum in practice we have to start the iterative procedure with different feasible initial values in order to compare the results reached in this way. A useful approach is to graph the function.

In some cases, it is reasonable to obtain a good initial value θ^0 for the vector of parameters through some computationally simple procedure, because the method (4.38) has good local convergence rates but not necessarily good convergence for points far away from the minimum. For example, an adequate initial value can be calculated ignoring the observation errors v_{yt} and v_{ut} defined in (3.1)-(3.2), and applying a consistent estimator to the reduced form (2.3).

4.7. Model Verification.

Once the estimate $\hat{\theta}$ for the parameter vector has been obtained, it is necessary to validate this process by verifying the model as follows. All of the assumptions made in the estimation process have to be tested.

As we have already shown, we obtained an exact expression for the information matrix during the estimation stage. The inverse of the information matrix gives us a lower bound for the covariance matrix of $\hat{\theta}$, which can be used in the process of verification to analyze the significance of the estimated parameters and of any possible overparametrization of the model by means of the correlation between these estimates.

Moreover, a fundamental element in the calculation of the likelihood function is the innovation process, $\tilde{z}_t$, defined by the Kalman filter. It should be checked that this process is white noise and a careful analysis of its autocorrelation and partial autocorrelation functions and of the cross-correlations function with the vector for the exogenous variables is needed. This can be done by traditional techniques in time series analysis, see Box and Jenkins (1976) and Newbold (1983).

In addition, the Lagrange-multiplier test, the Wald test and the likelihood-ratio test allow the specification of the estimated model to be analyzed. As is known, these tests are asymptotically equivalent, see Breusch and Pagan (1980) and Engle (1984). However, the Lagrange-multiplier test has the advantage that it only requires the estimation of the model under the null hypothesis. This test takes the following form:

$$LM = \left[\frac{\partial \ell(\hat{\theta}^r)}{\partial \theta}\right]' \left[M(\hat{\theta}^r)^{-1}\right] \left[\frac{\partial \ell(\hat{\theta}^r)}{\partial \theta}\right] \sim \chi_r^2 \qquad (4.39)$$

where $\hat{\theta}^r$ is the estimate of θ in the model with r restrictions, and $M(\hat{\theta}^r) = E\left[\frac{\partial^2 \ell(\hat{\theta}^r)}{\partial \theta \partial \theta'}\right]$. As we can see, only expressions that have already been calculated in the estimation process are used.

Again we would like to emphasize the importance of obtaining exact expressions for the gradient and for the information matrix. This applies both to the estimation of the model and to its verification.

In the verification process one may often have to choose between two or more models. Within the context of the maximum likelihood estimation we are working in, it is clear that insofar as we expand the model structure with more parameters, the minimum value of $\ell(\theta)$ decreases, since new degrees of freedom are added to the optimization problem.

Therefore, it seems reasonable to provide a criterion that offers a compromise between the minimum value of $\ell(\theta)$ and the number of parameters, p, to be estimated.

As we shall be seeing, this kind of criteria recently considered in the econometrics literature, are particular cases of the minimization of the General Criterion for Structural Selection (GCSS) defined as

$$GCSS = \frac{2}{N} \ell(\hat{\theta}) + \frac{\gamma(N)}{N} p + h(s) \qquad (4.40)$$

The first term on the right in (4.40) incorporates the minimization of $\ell(\theta)$, and it reflects the maximum likelihood criterion, the second term describes the estimated parameters cost. The last term h(s) is not always present and it discriminates among various input-output equivalent state-space forms.

As we have indicated, the expression (4.40) provides, for concrete values of $\gamma(N)$ and h(s), a series of model structure selection criteria based on theoretical information and coding theoretical ideas. Akaike (1974) was the first to propose a criterion of this kind, defined by

$\gamma(N) = 2$ and $h(s) = 0$. Therefore, Akaike's Information Criterion (AIC) has the form

$$\text{AIC} = \frac{2}{N} \ell(\hat{\theta}) + \frac{2}{N} p \qquad (4.41)$$

The minimization of (4.41) does not yield a consistent estimate of the model specification. The criterion (4.40) is said to be consistent if the probability of selecting a wrong number of parameters tends to zero as N tends to infinity.

Kashyap (1977) and Schwartz (1978), using Bayesian considerations, developed a criterion which gives consistent estimates of the model structure and that corresponds to:

$$\gamma(N) = \log N, \quad h(s) = 0 \qquad (4.42)$$

Hannan (1980) has also analyzed criteria of the form (4.40), and he showed that for

$$\gamma(N) \to \infty \quad \text{and} \quad \frac{\gamma(N)}{N} \to 0 \quad \text{as } N \to \infty \qquad (4.43)$$

with $h(s) = 0$, the specification minimizing GCSS is a consistent estimate of the true specification. There are many ways to satisfy (4.43), the values $\gamma(N) = \sqrt{N}$ and $\gamma(N) = \log N$ are two possibilities.

In order to decrease the risk of underfitting, Hannan (1980) also proposed

$$\gamma(N) = c \log \log N \quad \text{with } c > 2 \qquad (4.44)$$

which verifies (4.43) and still preserves the consistency property.

Finally, Rissanen (1983) proposed, based on coding theoretical principles, the Minimum Description Length (MDL) principle which corresponds to the following values in (4.40)

$$\gamma(N) = \log N, \quad h(s) = \frac{p}{N} \log(\hat{\theta}' \frac{1}{N} J\hat{\theta})$$

where $\frac{1}{N} J$ is the normalized hessian of the negative log-likelihood function.

The idea which supports the proposal of the MDL criterion is that of choosing the parameters in such a way that the model they define allows a redescription of the observed sequence with the smallest number of binary digits.

A fundamental stage in the specification and quality control of the model is assessing its consistency with economic theory. In relation to this matter, it is convenient to stress that *a priori* economic knowledge is primarily related to the long-run equilibrium solution of the system, since economic theory has very little to say about short-run behaviour. For this reason it is convenient to determine whether the equilibrium solutions of the estimated models are consistent with economic theory. Let us remember, see Chapter 2, that the long-run behaviour is characterized by the steady-state matrix multiplier $\Pi_\infty = \sum_{i=0}^{\infty} \Pi_i$, that in terms of the state-space formulation matrices is given by $\Pi_\infty = H(I - \Phi)^{-1}\Gamma + D$. The estimation algorithm proposed easily allows one to impose restrictions on Π_∞.

CHAPTER 5

EXTENSIONS OF THE ANALYSIS

The formulation of the econometric model with measurement errors given in Chapter 3 and defined by equations (3.17) to (3.26) can be extended in several directions. There is an immediate generalization following, for example, Harvey and Pierse (1984) and Ansley and Kohn (1983), for cases where observations are missing or where only specific temporal aggregates of the variables are observed. Also see the various studies compiled by Parzen (1984). Recall that the model in state-space form can be written:

$$x_{t+1} = \Phi x_t + E w_t \tag{5.1}$$

$$z_t = H x_t + C v_t \tag{5.2}$$

where $z_t = \begin{bmatrix} y_t^{*\prime} & u_t^{*\prime} \end{bmatrix}'$ is the vector formed by the endogenous and exogenous variables observed with error.

In Sections 5.1 and 5.2 we shall consider situations in which some of the observations are missing or aggregate. Then, in Section 5.3 we shall turn to situations where observation errors are correlated.

5.1. Missing Observations and Contemporaneous Aggregation.

If we assume that the observed variables are not the m components of the z_t vector, but rather the m^* components of the α_t vector, such that:

$$\alpha_t = H_t^* z_t \tag{5.3}$$

we will obtain for different values of $H_t^* \neq I_m$, the cases that interest us here.

For example, the value $H_t^* = 0$ corresponds to a situation where the m observations are missing for the time period t. The value $H_t^* = \begin{bmatrix} I_{m_1} & 0 \end{bmatrix}$ is a case where the last $(m - m_1)$ components of the vector z_t are not observed. Finally the expression:

$$H_t^* = \begin{bmatrix} i'_{m_1} & 0 \\ 0 & I_{m-m_1} \end{bmatrix}$$

where i'_{m_1} is a row vector of m_1 ones, corresponds to a situation where the sum of the first m_1 components of the vector z_t and the remaining $(m - m_1)$ components are observed.

If we substitute (5.3) into (5.2), we will have a new model to estimate, which includes the indicated special situations. This is given by:

$$x_{t+1} = \Phi x_t + E w_t \tag{5.4}$$

$$\alpha_t = \bar{H}_t x_t + \bar{C}_t v_t \tag{5.5}$$

where $\bar{H}_t = H_t^* H$ and $\bar{C}_t = H_t^* C$.

Note that in this case the matrices $\bar{H}_t$ and $\bar{C}_t$ are time varying. Thus the corresponding simplifications for a stationary situation of the Kalman filter cannot be applied. Chandrasekhar formulations can not be applied either.

The situation corresponding to moments in time at which all observations are missing, that is when $H_t^* = 0$, is equivalent to considering a value for the covariance matrix of observation errors of $R = rI$ with $r \gg 0$. This procedure amounts to not updating the estimates of the Kalman filter at such points in time, so that equations (4.8)-(4.12) reduce to

$$\hat{x}_{t+1|t} = \Phi \hat{x}_{t|t-1} + \Gamma u_t$$

$$P_{t+1|t} = \Phi P_{t|t-1}\Phi' + EQE'$$

Having thus defined the model to be estimated by (5.4) and (5.5) the function to be minimized, equivalent to expression (4.14), can be written in the form

$$\ell(\theta) = \sum_{t \in A} \left[\frac{m}{2} \log(2\pi) + \frac{1}{2} \log|B_t^\alpha| + \frac{1}{2} \tilde{\alpha}_t' (B_t^\alpha)^{-1} \tilde{\alpha}_t \right]$$

where the summation extends over the set A defined by the values of t ($1 \leq t \leq N$) such that $H_t^* \neq 0$. The values of $\tilde{\alpha}_t$ and B_t^α are given by the Kalman filter (4.8)-(4.12) with α_t, $\tilde{\alpha}_t$, B_t^α, $\bar{H}_t$ and $\bar{C}_t$ substituted for z_t, $\tilde{z}_t$, B_t, H and C, respectively.

In Section 4.4, we have made some remarks about the theoretical conditions for the identifiability of the model with equally-spaced data. In order to generalize these conditions for unequally-spaced data, a description of the process that generates the observations would be required. For example, a scalar MA(1) model is not identifiable if the available sample includes only observations for t = 1,3,5,7,... . Conditions for identifiability for unequally-spaced data have not yet been discussed with any generality.

If one wants to estimate the missing observations as well as estimate the model, the fixed-interval smoothing algorithm described in Appendix G should be employed.

5.2. Temporal Aggregation.

It is also possible to extend the formulation set out in Chapter 3 to a situation in which, for example, the intention is to estimate an econometric model corresponding to a specification which includes quarterly variables, when only annual sampling information is available. In this case it would be necessary to augment the state vector to include lagged values of the variables. Let us examine how we might proceed in this situation.

If y_t and u_t are flow variables, and the specifications given by (5.1) and (5.2) correspond to quarterly data, we could build the following model:

$$\begin{bmatrix} x_{t+1} \\ z_t \\ z_{t-1} \\ z_{t-2} \end{bmatrix} = \begin{bmatrix} \Phi & 0 & 0 & 0 \\ H & 0 & 0 & 0 \\ 0 & I & 0 & 0 \\ 0 & 0 & I & 0 \end{bmatrix} \begin{bmatrix} x_t \\ z_{t-1} \\ z_{t-2} \\ z_{t-3} \end{bmatrix} + \begin{bmatrix} E & 0 \\ 0 & C \\ 0 & 0 \\ 0 & 0 \end{bmatrix} \begin{bmatrix} w_t \\ v_t \end{bmatrix} \quad (5.6)$$

$$z_t^* = \begin{bmatrix} H & I & I & I \end{bmatrix} \begin{bmatrix} x_t \\ z_{t-1} \\ z_{t-2} \\ z_{t-3} \end{bmatrix} + Cv_t \quad (5.7)$$

where $t = 4, 8, 12, 16, \ldots$. In this formulation the endogenous and exogenous variables with measurement errors, given by the vector $z_t^* = \begin{bmatrix} y_t^{*\prime} & u_t^{*\prime} \end{bmatrix}'$, are the corresponding annual values.

If one is interested in obtaining estimates of the quarterly "observations" of the endogenous and exogenous variables as well, it is convenient to write the estimated model in the form

$$\begin{bmatrix} x_{t+1} \\ z_t \\ z_{t-1} \\ z_{t-2} \\ z_{t-3} \end{bmatrix} = \begin{bmatrix} \Phi & 0 & 0 & 0 & 0 \\ H & 0 & 0 & 0 & 0 \\ 0 & I & 0 & 0 & 0 \\ 0 & 0 & I & 0 & 0 \\ 0 & 0 & 0 & I & 0 \end{bmatrix} \begin{bmatrix} x_t \\ z_{t-1} \\ z_{t-2} \\ z_{t-3} \\ z_{t-4} \end{bmatrix} + \begin{bmatrix} E & 0 \\ 0 & C \\ 0 & 0 \\ 0 & 0 \\ 0 & 0 \end{bmatrix} \begin{bmatrix} w_t \\ v_t \end{bmatrix} \quad (5.8)$$

$$z_t^* = \begin{bmatrix} H & I & I & I & 0 \end{bmatrix} \begin{bmatrix} x_t \\ z_{t-1} \\ z_{t-2} \\ z_{t-3} \\ z_{t-4} \end{bmatrix} + Cv_t \quad (5.9)$$

The fixed-interval smoothing algorithm of Appendix G applied to (5.8)-(5.9) generates optimal estimates of the corresponding quarterly values of the variables. This procedure thus generalizes the interpolation methods used fundamentally in the econometrics literature, see for example, Chow and Lin (1976).

Obviously, if y_t and u_t were stock variables, this problem would reduce to one with missing observations, since stock variables are essentially continuous.

The extension to more general aggregation situations is straightforward.

Note that the treatment of the exogenous variables of the original model u_t as observed variables, according to the formulation set out in Chapter 3, makes it possible to consider temporal aggregates of the exogenous variables. Harvey and McKenzie (1984) consider only the possibility of temporal aggregation of the endogenous variables.

5.3. Correlated Measurement Errors.

It is also possible to extend the formulation to cases where the observation errors v_{yt} and v_{ut} are not white noise. In general, if these measurement errors follow the multivariate ARMA processes:

$$\phi_{v_y}(L) \, v_{yt} = \theta_{v_y}(L) \, a_{v_{yt}}$$

$$\phi_{v_u}(L) \, v_{ut} = \theta_{v_u}(L) \, a_{v_{ut}}$$

the corresponding state-space formulation will then be:

$$x_{t+1}^d = \Phi^d x_t^d + E^d w_t^d \tag{5.10}$$

$$v_{yt} = H^d x_t^d + C^d v_t^d \tag{5.11}$$

and

$$x^e_{t+1} = \Phi^e x^e_t + E^e w^e_t \tag{5.12}$$

$$v_{ut} = H^e x^e_t + C^e v^e_t \tag{5.13}$$

The matrices and vectors of (5.10)-(5.11) and of (5.12)-(5.13) are defined in a way similar to that employed for the ARMA model (3.13) given by (3.18)-(3.21).

If we write equations (3.25), (5.10) and (5.12), and equations (3.26), (5.11) and (5.13) in combined form we will obtain the following expressions, respectively:

$$\begin{bmatrix} x^a_{t+1} \\ x^b_{t+1} \\ x^d_{t+1} \\ x^e_{t+1} \end{bmatrix} = \begin{bmatrix} \Phi^a & \Gamma^a H^b & 0 & 0 \\ 0 & \Phi^b & 0 & 0 \\ 0 & 0 & \Phi^d & 0 \\ 0 & 0 & 0 & \Phi^e \end{bmatrix} \begin{bmatrix} x^a_t \\ x^b_t \\ x^d_t \\ x^e_t \end{bmatrix}$$

$$+ \begin{bmatrix} E^a & \Gamma^a C^b & 0 & 0 \\ 0 & E^b & 0 & 0 \\ 0 & 0 & E^d & 0 \\ 0 & 0 & 0 & E^e \end{bmatrix} \begin{bmatrix} \varepsilon_t \\ a_t \\ a_{v_{yt}} \\ a_{v_{ut}} \end{bmatrix} \tag{5.14}$$

$$\begin{bmatrix} y_t^* \\ u_t^* \end{bmatrix} = \begin{bmatrix} H^a & D^a H^b & H^d & 0 \\ 0 & H^b & 0 & H^e \end{bmatrix} \begin{bmatrix} x_t^a \\ x_t^b \\ x_t^d \\ x_t^e \end{bmatrix}$$

$$+ \begin{bmatrix} C^a & D^a C^b & C^d & 0 \\ 0 & C^b & 0 & C^e \end{bmatrix} \begin{bmatrix} \varepsilon_t \\ a_t \\ a_{v_{yt}} \\ a_{v_{ut}} \end{bmatrix} \quad (5.15)$$

These two equations represent the most general formulation for a dynamic econometric model having observation errors in the endogenous and exogenous variables in such a way that these errors follow multivariate ARMA processes.

CHAPTER 6

NUMERICAL RESULTS

In this chapter we shall concentrate on an analysis of the estimation problems for the different parametrizations of a model corresponding to a dynamic specification developed naturally under assumptions frequently found in economic theory, such as partial adjustment.

This model was recently examined by Aigner et al.(1984, pp. 1380-1386) to analyze problems of measurement errors in the variables and is given as follows:

$$y_t + \beta y_{t-1} + \gamma u_t = \varepsilon_t \,, \quad \varepsilon_t \sim N(0,\sigma_\varepsilon^2) \tag{6.1}$$

$$y_t^* = y_t + v_{yt} \tag{6.2}$$

$$u_t^* = u_t + v_{ut} \tag{6.3}$$

Both endogenous and exogenous variables are observed with error. These errors are characterized by:

$$v_{yt} \sim N(0,\sigma_{v_y}^2) \,, \quad v_{ut} \sim N(0,\sigma_{v_u}^2) \tag{6.4}$$

In line with what we established in Chapter 3, we will assume that the exogenous variable follows an AR(1) process, so that:

$$u_t + \phi u_{t-1} = a_t \,, \quad a_t \sim N(0,\sigma_a^2) \tag{6.5}$$

Thus we can write the previous equations, in the light of (3.25)-(3.26), as follows:

$$\begin{bmatrix} x_{t+1}^a \\ x_{t+1}^b \end{bmatrix} = \begin{bmatrix} -\beta & \beta\gamma \\ 0 & -\phi \end{bmatrix} \begin{bmatrix} x_t^a \\ x_t^b \end{bmatrix} + \begin{bmatrix} -\beta & \beta\gamma \\ 0 & -\phi \end{bmatrix} \begin{bmatrix} \varepsilon_t \\ a_t \end{bmatrix} \tag{6.6}$$

$$\begin{bmatrix} y_t^* \\ x_t^* \end{bmatrix} = \begin{bmatrix} 1 & -\gamma \\ 0 & 1 \end{bmatrix} \begin{bmatrix} x_t^a \\ x_t^b \end{bmatrix} + \begin{bmatrix} 1 & -\gamma & 1 & 0 \\ 0 & 1 & 0 & 0 \end{bmatrix} \begin{bmatrix} \varepsilon_t \\ a_t \\ v_{yt} \\ v_{ut} \end{bmatrix} \quad (6.7)$$

The model we have estimated then is the one given by the two last equations where the covariance matrices for the white noises are:

$$Q = \begin{bmatrix} \sigma_\varepsilon^2 & 0 \\ 0 & \sigma_a^2 \end{bmatrix}, \quad S = \begin{bmatrix} \sigma_\varepsilon^2 & 0 & 0 & 0 \\ 0 & \sigma_a^2 & 0 & 0 \end{bmatrix}$$

$$R = \begin{bmatrix} \sigma_\varepsilon^2 & 0 & 0 & 0 \\ 0 & \sigma_a^2 & 0 & 0 \\ 0 & 0 & \sigma_{v_y}^2 & 0 \\ 0 & 0 & 0 & \sigma_{v_u}^2 \end{bmatrix}$$
(6.8)

We have assumed that ε_t, a_t, v_{yt} and v_{ut} are independent.

Note that in this problem, n = m, and that the Ricatti matrix equation (A.16) has been used in estimation, because the alternative formulations of the Chandrasekhar type offer no computational advantages whatever.

For the purpose of analyzing the effects of autocorrelation in the structure of the exogenous variable, we will also consider the previous model in a situation where $\phi = 0$, that is, when the exogenous variable follows a white noise process given by:

$$u_t \sim N(0, \sigma_a^2) \quad (6.9)$$

In this situation, the parametrization given by (6.6) and (6.7) reduces to:

$$x^a_{t+1} = -\beta x^a_t - \beta \varepsilon_t + \beta\gamma a_t$$

$$y^*_t = x^a_t + \varepsilon_t - \gamma a_t + v_{yt}$$

$$u^*_t = a_t + v_{ut}$$

The last equation shows that only the sum $\sigma^2_a + \sigma^2_{v_u}$ is identifiable. The two first equations clearly show that in this situation β is the only identified parameter.

Thus, to make the specification identifiable for a situation in which the exogenous variable follows a white noise process, we assume that this variable is observed without error. It is not convenient here to include the exogenous variable among the observed variables of the model in state-space as we have suggested in Chapter 3. The state-space formulation would then be:

$$x_{t+1} = -\beta x_t + \beta\gamma u_t - \beta\varepsilon_t \tag{6.10}$$

$$y^*_t = x_t - \gamma u_t + \varepsilon_t + v_{yt} \tag{6.11}$$

Note that with the notation used in (4.1) and (4.2) this situation corresponds to one where $\Gamma = \beta\gamma$, $D = -\gamma$, rather than to a formulation where the exogenous variables are observed with error, in which case: $\Gamma = 0$, $D = 0$. We have estimated the parametrization of (6.10)-(6.11) to compare this result with the one in which the exogenous variable observed with error follows an AR(1) process.

To compare results obtained when the exogenous variable follows an AR(1) process and when it is white noise, we have to define the following signal-to-noise ratios:

$$\mu_y = \frac{\text{var}(y_t)}{\text{var}(v_{yt})}, \quad \mu_u = \frac{\text{var}(u_t)}{\text{var}(v_{ut})} \tag{6.12}$$

It is then easy to show, on the basis of previous specifications, that

$$\mu_y = \frac{1}{\sigma_{v_y}^2} \frac{1}{1-\beta^2} \left[\sigma_\varepsilon^2 + \frac{\gamma^2 \sigma_a^2}{1-\phi^2} \frac{1+\beta\phi}{1-\beta\phi} \right] \tag{6.13}$$

$$\mu_u = \frac{1}{\sigma_{v_u}^2} \frac{\sigma_a^2}{1-\phi^2} \tag{6.14}$$

In all cases we have assumed the same levels of error in the observations of y_t and u_t given, in succession, by:

$$\sigma_{v_y}^2 = \sigma_{v_u}^2 = 0.01, \quad \sigma_{v_y}^2 = \sigma_{v_u}^2 = 0.1, \quad \sigma_{v_y}^2 = \sigma_{v_u}^2 = 0.5,$$

$$\sigma_{v_y}^2 = \sigma_{v_u}^2 = 1.0 \tag{6.15}$$

In addition we have assumed:

$$\beta = -0.5, \quad \gamma = -0.7, \quad \phi = -0.8 \tag{6.16}$$

Whenever we have employed the assumption that u_t is white noise, we have assumed the following values:

$$\sigma_a^2 = 1.0, \quad \sigma_\varepsilon^2 = 1.0 \tag{6.17}$$

These values give the following signal-to-noise ratios:

$$\mu_y = 199, 20, 4, 2 \tag{6.18}$$

$$\mu_u = 100, 10, 2, 1 \tag{6.19}$$

So if we want the signal-to-noise ratios, μ_y and μ_u, to be the same when u_t is white noise and when it follows an AR(1) process, we would have to take the following values for the AR(1) situation:

$$\sigma_a^2 = 0.36, \quad \sigma_\varepsilon^2 = 0.35 \tag{6.20}$$

These values are obtained using equations (6.13) and (6.14), once the values given in (6.15), (6.16), (6.18) and (6.19) are fixed.

We see then, from (6.18) and (6.19), that we have reached situations where we are considering noise levels amounting to 50% of the endogenous variable, and 100% of the exogenous variable.

Using the theoretical values of the parameters we generated 20 realizations of 500 observations for each situation we examined. The results of the estimation process, assuming u_t to be white noise, are given in Table 1 (p. 64). The results for the case in which the exogenous variable follows an AR(1) process are given in Table 2 (p. 65).

In all cases we have also estimated the following model:

$$y_t^* + \beta^* y_{t-1}^* + \gamma^* u_t^* = \varepsilon_t^* \tag{6.21}$$

This corresponds to the specification where the variables are measured without error, but where we use sample information containing observation errors.

Where we assume that the exogenous variable is white noise and it is observed without error, we have also estimated:

$$y_t^* + \beta y_{t-1}^* + \gamma u_t = \eta_t + \theta \eta_{t-1} \tag{6.22}$$

which is the specification for the endogenous variable observed with error.

We repeated all the indicated estimations using only the first 100 observations of each realization. The results are shown in Tables 3 and 4 (pp. 66 and 67).

We also estimated the basic model with the exogenous variable with an AR(1) structure having values of ϕ near zero, in order to analyze the behaviour of the algorithm given in (4.38). In fact the Monte-Carlo analysis was done with $\phi = 0.1$ and $\phi = 0.2$, which imply situations approaching an unidentifiable model. The updating procedure given by

(4.38) yields very acceptable results, comparable to those shown here, which we shall comment on next.

Although Tables 1, 2, 3 and 4 are in themselves self explanatory, it is worth bearing the following in mind:

- The estimates are in all cases acceptable. The estimation situation deteriorates as the magnitude of the observation errors increases. The mean values are very close to their true values and the confidence intervals contain the true values. Moreover, the confidence intervals largely coincide with the theoretical values derived from the exact expression of the information matrix, and differ in many cases from the values obtained through the approximate expression for the information matrix given by (4.29).

- The estimates are clearly better when the exogenous variable follows an AR(1) process, despite the fact that in this situation the exogenous variable is measured with error. It then follows that, through the information introduced by the autocorrelation structure, the effect of uncertainty as a result of observation error in the exogenous variable is offset.

- When the presence of observation errors is overlooked in the process of estimation with the exogenous variable observed with error, the bias in the estimates of β and γ is significant, even when the signal-to-noise ratio is high. This bias is especially relevant in the case of the parameter γ.

- The analysis of the eigenvalues and eigenvectors of the information matrix suggests that, in general and in the identified situations, it is well-behaved. Nevertheless, the value of $\tau = \lambda_{max}/\lambda_{min}$ is, in certain cases, fairly sensitive to the size of the sample. For example, when $\sigma^2_{v_y} = \sigma^2_{v_u} = 0.1$ with 100 observations, the greatest value for τ was approximately 100. In the case of 500 observations and in the same situation, this value was around 10.

- The mean CPU time used for the estimation of the previous models in a UNIVAC 1100, in the state-space formulation, was 110 seconds for Table 1, 180 seconds for Table 2, 30 seconds for Table 3, and 50 seconds for Table 4. Though these times depend, among other factors, on the initial values θ^0 used in the minimization algorithm, they serve as an indication of the computational burden that the estimation of these models implies. The initial conditions in our analysis were always within the interval of $\theta^* \pm 0.3\theta^*$, θ^* being the theoretical value of the parameters.

- It is important to use the exact analytical expression for the information matrix. In fact, empirical experience shows that the approximation used by Watson and Engle (1983) can differ substantially from the true values, which can lead to errors in the statistical verification of the model. Such discrepancy is even greater with the values given for the approximations of the hessian used in the standard optimization algorithms such as the Davidon-Fletcher-Powell or Marquardt procedures. If one wants to avoid the calculation of the exact expression for the information matrix in every iteration of the optimization process, a satisfactory solution might be to use algorithms which require only information about the gradient. Once the optimum is achieved the information matrix could be evaluated by its exact expression.

To summarize, the numerical results show that the estimation procedure proposed for econometric models with measurement errors is fully satisfactory. Besides the results specified here and collected in Tables 1, 2, 3 and 4, analyses have been carried out, for N = 100 and N = 500, with the basic formulation (6.1)-(6.3) and under the assumption that 25% of the observations are missing in a random way. These analyses have given analogous results to the ones discussed here, and conclusions similar to the ones shown here can be reached from it.

The general scope of application of this method, as we have discussed in Chapters 3, 4, and 5, contrasts with the limitations of the procedures suggested in the econometrics literature for cases where the exogenous variable is observed with error. These procedures are developed in the area of spectral analysis, and assume that the observation errors are present only at certain frequencies, see Aigner et al.(1984).

In their recent work on the current state of this problem, Aigner et al.(1984, pp. 1380-1386) conclude that the calculation of efficient estimates for dynamic models with measurement errors is a formidable task, and that there is today no simple computational procedure for the treatment of this problem.

These conclusions are reached in the context of their analysis of the model estimated here, and defined by equations (6.1)-(6.4). On the basis of the results shown here, we believe that the procedures suggested in this work cover the existing gap in current methodology for the estimation of econometric models with measurement errors and missing observations in both the endogenous and exogenous variables.

	$y_t + \beta y_{t-1} + \gamma u_t = \epsilon_t$, $y_t^* = y_t + v_{yt}$			$y_t^* + \beta^* y_{t-1}^* + \gamma^* u_t^* = \epsilon_t^*$			$y_t^* + \beta y_{t-1}^* + \gamma u_t = \eta_t + \theta \eta_{t-1}$			
$\sigma_{v_y}^2$	$\hat{\beta}$	$\hat{\gamma}$	$\hat{\sigma}_\epsilon^2$	$\hat{\sigma}_{v_y}^2$	$\hat{\beta}^*$	$\hat{\gamma}^*$	$\hat{\sigma}_{\epsilon^*}^2$	$\hat{\beta}$	$\hat{\gamma}$	$\hat{\theta}$	$\hat{\sigma}_\eta^2$
0.01	−0.51 (0.03)	−0.70 (0.04)	0.94 (0.11)	0.05 (0.09)	−0.50 (0.04)	−0.71 (0.04)	1.01 (0.01)	−0.51 (0.03)	−0.70 (0.04)	−0.02 (0.06)	1.00 (0.01)
0.10	−0.51 (0.03)	−0.70 (0.04)	0.95 (0.14)	0.13 (0.12)	−0.48 (0.04)	−0.71 (0.04)	1.12 (0.03)	−0.52 (0.04)	−0.70 (0.04)	−0.06 (0.06)	1.12 (0.03)
0.50	−0.52 (0.05)	−0.70 (0.05)	0.94 (0.20)	0.55 (0.20)	−0.41 (0.04)	−0.71 (0.05)	1.60 (0.08)	−0.52 (0.05)	−0.70 (0.05)	−0.18 (0.08)	1.12 (0.08)
1.00	−0.52 (0.06)	−0.70 (0.06)	0.93 (0.27)	1.05 (0.28)	−0.34 (0.05)	−0.71 (0.06)	2.17 (0.12)	−0.52 (0.06)	−0.70 (0.06)	−0.26 (0.09)	2.13 (0.11)

Table 1. Results of estimating with 500 observations and $u_t \sim N(0, \sigma_a^2)$.
Theoreticcal values: $\beta = -0.5$, $\gamma = -0.7$, $\sigma_\epsilon^2 = 1.0$, $\sigma_a^2 = 1.0$

	$y_t + \beta y_{t-1} + \gamma u_t = \epsilon_t$, $u_t + \phi u_{t-1} = a_t$					$y^*_t = y_t + v_{yt}$, $u^*_t = u_t + v_{ut}$		$y^*_t + \beta^* y^*_{t-1} + \gamma^* u^*_t = \epsilon^*_t$		
$\sigma^2_{v_y} = \sigma^2_{v_u}$	$\hat\beta$	$\hat\gamma$	$\hat\sigma^2_\epsilon$	$\hat\phi$	$\hat\sigma^2_a$	$\hat\sigma^2_{v_y}$	$\hat\sigma^2_{v_u}$	$\hat\beta^*$	$\hat\gamma^*$	$\hat\sigma^2_{\epsilon^*}$
0.01	-0.50 (0.02)	-0.70 (0.02)	0.33 (0.04)	-0.80 (0.02)	0.36 (0.02)	0.02 (0.03)	0.01 (0.01)	-0.50 (0.03)	-0.70 (0.03)	0.37 (0.01)
0.10	-0.50 (0.03)	-0.70 (0.05)	0.33 (0.06)	-0.80 (0.03)	0.36 (0.04)	0.12 (0.06)	0.10 (0.03)	-0.49 (0.03)	-0.63 (0.04)	0.52 (0.03)
0.50	-0.49 (0.06)	-0.71 (0.10)	0.33 (0.12)	-0.80 (0.03)	0.37 (0.07)	0.52 (0.13)	0.50 (0.06)	-0.47 (0.03)	-0.48 (0.04)	1.14 (0.07)
1.00	-0.48 (0.09)	-0.72 (0.15)	0.34 (0.18)	-0.79 (0.03)	0.37 (0.09)	1.01 (0.20)	1.00 (0.10)	-0.42 (0.03)	-0.39 (0.05)	1.85 (0.12)

Table 2. Results of estimating with 500 observations and $u_t + \phi u_{t-1} = a_t$.
Theoreticcal values: $\beta = -0.5$, $\gamma = -0.7$, $\sigma^2_\epsilon = 0.35$, $\phi = -0.8$, $\sigma^2_a = 0.36$

	$y_t + \beta y_{t-1} + \gamma u_t = \epsilon_t$, $y_t^* = y_t + v_{yt}$				$y_t^* + \beta^* y_{t-1}^* + \gamma^* u_t^* = \epsilon_t^*$				$y_t^* + \beta y_{t-1}^* + \gamma u_t = \eta_t + \theta \eta_{t-1}$			
$\sigma_{v_y}^2$	$\hat{\beta}$	$\hat{\gamma}$	$\hat{\sigma}_\epsilon^2$	$\hat{\sigma}_{v_y}^2$	$\hat{\beta}^*$	$\hat{\gamma}^*$	$\hat{\sigma}_{\epsilon^*}^2$		$\hat{\beta}$	$\hat{\gamma}$	$\hat{\theta}$	$\hat{\sigma}_\eta^2$
0.01	−0.51 (0.08)	−0.69 (0.08)	0.89 (0.12)	0.04 (0.07)	−0.51 (0.09)	−0.69 (0.08)	0.93 (0.10)		−0.50 (0.11)	−0.69 (0.08)	−0.03 (0.11)	0.93 (0.10)
0.10	−0.50 (0.09)	−0.70 (0.09)	0.91 (0.17)	0.08 (0.09)	−0.49 (0.09)	−0.70 (0.09)	1.04 (0.14)		−0.49 (0.12)	−0.70 (0.09)	−0.01 (0.13)	1.03 (0.14)
0.50	−0.49 (0.11)	−0.72 (0.11)	1.06 (0.30)	0.34 (0.23)	−0.42 (0.09)	−0.72 (0.12)	1.47 (0.21)		−0.48 (0.15)	−0.72 (0.12)	−0.11 (0.16)	1.45 (0.20)
1.00	−0.47 (0.12)	−0.74 (0.13)	1.17 (0.45)	0.70 (0.41)	−0.35 (0.10)	−0.73 (0.15)	1.98 (0.26)		−0.47 (0.18)	−0.74 (0.14)	−0.19 (0.20)	1.94 (0.25)

Table 3. Results of estimating with 100 observations and $u_t \sim N(0, \sigma_a^2)$.
Theoreticcal values: $\beta = -0.5$, $\gamma = -0.7$, $\sigma_\epsilon^2 = 1.0$, $\sigma_a^2 = 1.0$

$\sigma^2_{v_y} = \sigma^2_{v_u}$	$y_t + \beta y_{t-1} + \gamma u_t = \epsilon_t$, $u_t + \phi u_{t-1} = a_t$,				$y^*_t = y_t + v_{yt}$ $u^*_t = u_t + v_{ut}$			$y^*_t + \beta^* y^*_{t-1} + \gamma^* u^*_t = \epsilon^*_t$		
	$\hat{\beta}$	$\hat{\gamma}$	$\hat{\sigma}^2_\epsilon$	$\hat{\phi}$	$\hat{\sigma}^2_a$	$\hat{\sigma}^2_{v_y}$	$\hat{\sigma}^2_{v_u}$	$\hat{\beta}^*$	$\hat{\gamma}^*$	$\hat{\sigma}^{2*}_\epsilon$
0.01	−0.52 (0.06)	−0.70 (0.08)	0.29 (0.07)	−0.78 (0.07)	0.32 (0.08)	0.04 (0.05)	0.02 (0.04)	−0.50 (0.06)	−0.71 (0.07)	0.38 (0.02)
0.10	−0.51 (0.09)	−0.71 (0.13)	0.29 (0.13)	−0.75 (0.08)	0.37 (0.10)	0.13 (0.09)	0.09 (0.08)	−0.49 (0.07)	−0.62 (0.08)	0.54 (0.07)
0.50	−0.51 (0.10)	−0.72 (0.11)	0.33 (0.35)	−0.75 (0.09)	0.38 (0.16)	0.56 (0.33)	0.48 (0.19)	−0.46 (0.07)	−0.46 (0.09)	1.17 (0.15)
1.00	−0.50 (0.12)	−0.75 (0.12)	0.35 (0.45)	−0.76 (0.10)	0.41 (0.26)	0.99 (0.53)	0.92 (0.32)	−0.41 (0.08)	−0.39 (0.11)	1.91 (0.26)

Table 4. Results of estimating with 100 observations and $u_t + \phi u_{t-1} = a_t$. Theoreticcal values: $\beta = -0.5$, $\gamma = -0.7$, $\sigma^2_\epsilon = 0.35$, $\phi = -0.8$, $\sigma^2_a = 0.36$

CHAPTER 7

CONCLUSIONS

In this monograph we have developed a new formulation for dynamic econometric models with measurement errors. We have also obtained an algorithm for the maximum likelihood estimation of all the parameters of the model. The estimates obtained in this way are consistent and asymptotically normal and efficient.

We have applied this methodology to the estimation of different parametrizations of a model used frequently in the econometric literature, and for which existing estimation procedures do not offer acceptable results. The numerical results obtained by the proposed methodology are fully satisfactory.

The main contributions of our work are:

- A minimal dimensional state-space formulation for econometric models with measurement errors in both endogenous and exogenous variables.

- Analytical expressions for the maximization of the exact likelihood function of the model. Our estimates include the components of the initial state that are identifiable. The type of initial conditions proposed is just as suitable for use in the Kalman filter Riccati equations as in those of the Chandrasekhar type which frequently offer appreciable computational advantages.

- An exact expression for the information matrix. This result is particularly relevant to both the iterative process of maximization of the likelihood function and to the validity of the tests for the verification of the model.

- A generalization of the results for situations in which observations are missing, where only contemporaneous or temporal aggregates of the variables are observed, and where the measurement errors are correlated.

What is proposed, and has been numerically tested, is a general procedure for the maximum likelihood estimation of dynamic econometric models with measurement errors and missing observations in both the endogenous and exogenous variables.

The procedure described here is applicable to a series of problems, recently dealt with in the econometrics literature, such as estimation of models with rational expectations, seasonal adjustment of time series, interpolation in time series, estimation of models with composite moving average disturbance terms, and to the analysis of time series with random walk and other nonstationary components.

APPENDIX A

KALMAN FILTER AND CHANDRASEKHAR EQUATIONS

A.1. Kalman Filter.

Let us suppose the system is given by

$$x_{t+1} = \Phi x_t + \Gamma u_t + E w_t \qquad (A.1)$$

$$z_t = H x_t + D u_t + C v_t \qquad (A.2)$$

such that x_0 has a mean value $\bar{x}_0$ and a covariance matrix P_0 and where w_t and v_t are white noise processes with

$$E[w_t] = 0, \quad E[v_t] = 0$$

$$E\left\{\begin{bmatrix} w_{t_1} \\ v_{t_1} \end{bmatrix} \begin{bmatrix} w'_{t_2} & v'_{t_2} \end{bmatrix}\right\} = \begin{bmatrix} Q & S \\ S' & R \end{bmatrix} \delta_{t_1 t_2} \qquad (A.3)$$

and $Q \geq 0$ and $R > 0$.

The initial state x_0 and the perturbations w_t and v_t have normal distributions.

If we use the notation:

$$\hat{x}_{i|j} = E[x_i | z^j] \qquad (A.4)$$

$$P_{i|j} = E\left\{[x_i - \hat{x}_{i|j}][x_i - \hat{x}_{i|j}]' | z^j\right\} \qquad (A.5)$$

where $z^j = \{z'_1, \ldots, z'_j\}'$, the Kalman filter will be given by the following expressions:

$$\hat{x}_{t+1|t} = \Phi\hat{x}_{t|t-1} + \Gamma u_t + K_t\left[z_t - H\hat{x}_{t|t-1} - Du_t\right] \qquad (A.6)$$

$$\hat{x}_{0|-1} = E\left[x_0\right] = \bar{x}_0 \qquad (A.7)$$

$$K_t = \left[\Phi P_{t|t-1}H' + ESC'\right]\left[HP_{t|t-1}H' + CRC'\right]^{-1} \qquad (A.8)$$

$$P_{t+1|t} = \Phi P_{t|t-1}\Phi' + EQE' - K_t\left[HP_{t|t-1}H' + CRC'\right]K_t' \qquad (A.9)$$

$$P_{0|-1} = P_0 \qquad (A.10)$$

The formulation of the Kalman filter given by (A.6)-(A.10) is obtained by the direct application of the following results.

If X is a random variable such that $X \sim N(\nu, \Omega)$, and we define $Y = AX + b$, then:

$$Y \sim N(A\nu + b, A\Omega A') \qquad (A.11)$$

In addition, if we assume that $\mu = A\nu + b$ and $\Sigma = A\Omega A'$, and that:

$$Y = \begin{bmatrix} Y_1 \\ Y_2 \end{bmatrix}, \quad \mu = \begin{bmatrix} \mu_1 \\ \mu_2 \end{bmatrix}, \quad \Sigma = \begin{bmatrix} \Sigma_{11} & \Sigma_{12} \\ \Sigma_{12}' & \Sigma_{22} \end{bmatrix}$$

it can be shown that the conditional distribution for Y_1 given $Y_2 = y_2$ is:

$$N\left[\mu_1 + \Sigma_{12}\Sigma_{22}^{-1}(y_2 - \mu_2), \Sigma_{11} - \Sigma_{12}\Sigma_{22}^{-1}\Sigma_{12}'\right] \qquad (A.12)$$

As a result, if we re-write (A.1) and (A.2) in condensed state-space form:

$$\begin{bmatrix} x_{t+1} \\ z_t \end{bmatrix} = \begin{bmatrix} \Phi & E & 0 \\ H & 0 & C \end{bmatrix}\begin{bmatrix} x_t \\ w_t \\ v_t \end{bmatrix} + \begin{bmatrix} \Gamma \\ D \end{bmatrix}u_t \qquad (A.13)$$

the joint distribution of $\begin{bmatrix} x_{t+1} \\ z_t \end{bmatrix}$, given z^{t-1}, is normal with mean:

$$\begin{bmatrix} \Phi\hat{x}_{t|t-1} \\ H\hat{x}_{t|t-1} \end{bmatrix} + \begin{bmatrix} \Gamma \\ D \end{bmatrix} u_t \tag{A.14}$$

and covariance:

$$\begin{bmatrix} \Phi P_{t|t-1}\Phi' + EQE' & \Phi P_{t|t-1}H' + ESC' \\ HP_{t|t-1}\Phi' + CS'E' & HP_{t|t-1}H' + CRC' \end{bmatrix} \tag{A.15}$$

This is the direct consequence of (A.11).

Expressions (A.6), (A.8), and (A.9) for the Kalman filter are obtained directly using (A.14) and (A.15) in the result given by (A.12).

On the basis of (A.1), (A.2) and (A.6), we can write

$$\tilde{x}_{t+1|t} = \bar{\Phi}_t \tilde{x}_{t|t-1} + Ew_t - K_t Cv_t$$

where

$$\tilde{x}_{t+1|t} = x_{t+1} - \hat{x}_{t+1|t}$$

$$\bar{\Phi}_t = \Phi - K_t H$$

Therefore

$$P_{t+1|t} = \bar{\Phi}_t P_{t|t-1} \bar{\Phi}_t' + \begin{bmatrix} E & -K_t C \end{bmatrix} \begin{bmatrix} Q & S \\ S' & R \end{bmatrix} \begin{bmatrix} E' \\ -C'K_t' \end{bmatrix} \tag{A.16}$$

or

$$P_{t+1|t} = \bar{\Phi}_t P_{t|t-1} \bar{\Phi}'_t + EQE' + K_t CRC' K'_t - K_t CS'E' - ESC'K'_t \qquad (A.17)$$

Expression (A.16) reveals that $P_{t+1|t}$ is the result of the sum of two positive semidefinite matrices. This does not occur in (A.9). Consequently, numerical computations based upon (A.16) will be better conditioned than those based upon (A.9). It is important to observe that possible computational errors in the calculation of K_t by means of (A.8) imply first-order errors in $P_{t+1|t}$ when (A.9) is used, and second-order errors when (A.16) is used.

Note that the expression for the Kalman filter we have obtained is different from that normally used in the econometrics literature, because we have taken into account the possibility that the noise terms for the system and the observation equation may be correlated. We have also included in this formulation the matrix C for the noise distribution in the observation equation.

The possible existence of contemporaneous correlation between w_t and v_t, characterized by matrix S, is not frequent in the Kalman filter formulations used in the econometrics literature, and leads to the terms $K_t CS'E'$ and $ESC'K'_t$ in (A.17). The inclusion of this effect could also be taken into account with a slight redefinition of the state vector, see Anderson and Moore (1979, p. 104).

For a more extensive treatment of the Kalman filter, the following may be consulted: Jazwinsky (1970) and Anderson and Moore (1979).

A.2. **Chandrasekhar Equations**.

The expressions for the Kalman filter given by equations (A.6) to (A.10) are valid when the system matrices are time-varying. However, it is possible to obtain alternative formulations when the matrices are not time-varying which, in some cases, are computationally more efficient.

If we write the matrix equations for the Kalman filter in the following form:

$$B_t = HP_{t|t-1}H' + CRC' \qquad (A.18)$$

$$K_t = \left[\Phi P_{t|t-1}H' + ESC'\right] B_t^{-1} \qquad (A.19)$$

$$P_{t+1|t} = \Phi P_{t|t-1}\Phi' + EQE' - K_t B_t K_t' \qquad (A.20)$$

it will be possible to obtain new algorithms which are computationally efficient when n >> m. This can be done by substituting for the Riccati equation (A.20) a group of difference equations of the Chandrasekhar type, see Morf et al. (1974).

One of the forms of this algorithm is as follows:

$$B_t = B_{t-1} + HY_{t-1}X_{t-1}Y_{t-1}'H' \qquad (A.21)$$

$$K_t = \left[K_{t-1}B_{t-1} + \Phi Y_{t-1}X_{t-1}Y_{t-1}'H'\right] B_t^{-1} \qquad (A.22)$$

$$Y_t = \left[\Phi - K_t H\right] Y_{t-1} \qquad (A.23)$$

$$X_t = X_{t-1} + X_{t-1}Y_{t-1}'H'B_{t-1}^{-1}HY_{t-1}X_{t-1} \qquad (A.24)$$

where the X_t and Y_t matrices have dimensions ($\alpha \times \alpha$) and ($n \times \alpha$) respectively. The value of α will depend on the specific properties of the system set out in (A.1) and (A.2), and to which we shall refer later.

We shall go on to see that in certain cases it is useful to use the following expressions instead of (A.23) and (A.24):

$$Y_t = \left[\Phi - K_{t-1}H\right] Y_{t-1} \qquad (A.23')$$

$$X_t = X_{t-1} - X_{t-1}Y_{t-1}'H'B_t^{-1}HY_{t-1}X_{t-1} \qquad (A.24')$$

In the final part of this appendix we shall obtain all of these expressions.

The initial conditions of (A.21) and (A.22) will be given by:

$$B_0 = HP_0H' + CRC' \tag{A.25}$$

$$K_0 = \left[\Phi P_0 H' + ESC'\right] B_0^{-1} \tag{A.26}$$

To obtain the initial conditions for (A.23) and (A.24), the expression

$$\delta P_0 = \Phi P_0 \Phi' + EQE' - K_0 B_0 K_0' - P_0 \tag{A.27}$$

will have to be factorized in the following form:

$$\delta P_0 = \overline{Y}\overline{X}\overline{Y}' \tag{A.28}$$

where $\overline{X}$ and $\overline{Y}$ are ($\alpha \times \alpha$) and ($n \times \alpha$) matrices, and

$$\overline{X} = \begin{bmatrix} X_+ & 0 \\ 0 & X_- \end{bmatrix} \quad \text{where } X_+ > 0, \ X_- < 0$$

With this factorization, which is not unique, the initial conditions of (A.23) and (A.24) will be:

$$X_0 = \overline{X}, \quad Y_0 = \overline{Y} \tag{A.29}$$

The value of α will be determined by the initial conditions because

$$\alpha = \text{rank}(\delta P_0) \tag{A.30}$$

Note that the calculation of $P_{t+1|t}$ is not necessary for the proposed algorithm, but it can be calculated by means of

$$P_{t+1|t} = P_{t|t-1} + Y_t X_t Y_t' \tag{A.31}$$

Since the evaluation of the negative log-likelihood function to be minimized given by (4.14) does not include $P_{t+1|t}$, the algorithm (A.21)-(A.24) can be used for the maximum likelihood estimation of the model.

Note that the expression (A.24) is a homogeneous Riccati equation, and therefore can be converted to a linear equation in the variable X_t^{-1}. As the X_t matrix is ($\alpha \times \alpha$), while $P_{t+1|t}$ is ($n \times n$), we will obtain important computational advantages when $\alpha \ll n$. Moreover, unlike $P_{t+1|t}$, the X_t matrix need not to be positive semidefinite.

In our problem, as we have seen in Section 4.3, there are three types of possible initial conditions. Each of them, as we shall see below, yields values for α which make the algorithm computationally attractive.

In the case of a stationary model, the value of P_0 will be given by the solution of the Lyapunov equation:

$$P = \Phi P \Phi' + EQE' \qquad (A.32)$$

Thus expression (A.27) will be:

$$\delta P_0 = - \left[\Phi PH' + ESC'\right]\left[HPH' + CRC'\right]^{-1}\left[\Phi PH' + ESC'\right]' \leq 0 \qquad (A.33)$$

Assuming H has full rank, we have that $\alpha = \text{rank}\left[\delta(P_0)\right] \leq \min(n,m)$ and an obvious factorization of (A.33) will give the following initial conditions:

$$X_0 = - \left[HPH' + CRC'\right]^{-1} < 0 \qquad (A.34)$$

$$Y_0 = \Phi PH' + ESC' \qquad (A.35)$$

We can then use (A.34) and (A.35) instead of (A.29), and if we want to assure the negative definite property and the monotone nonincreasing propagation of X_t, we should use expressions (A.23') and (A.24') instead of (A.23) and (A.24).

When the model is nonstationary, we will consider the initial state x_0 as being a constant but unknown parameter, so that $P_0 = 0$. In this case expression (A.27) becomes

$$\delta P_0 = E\left[Q - SC'(CRC')^{-1}CS'\right]E' \qquad (A.36)$$

Assuming E has full rank, we have that

$$\alpha = \text{rank}\left[\delta(P_0)\right] \leq \min(n, n_w) \qquad (A.37)$$

where n_w is the dimension of w_t.

One obvious factorization of (A.36) is:

$$X_0 = Q - SC'\left[CRC'\right]^{-1}CS' \qquad (A.38)$$

$$Y_0 = E \qquad (A.39)$$

The initial conditions will be, in this case, those given by (A.38) and (A.39), together with

$$B_0 = CRC' \qquad (A.40)$$

$$K_0 = ESC'\left[CRC'\right]^{-1} \qquad (A.41)$$

Given $X_0 > 0$, the use of (A.24) assures the monotone non-decreasing character of X_t. Taking (A.31) into account, we can also assure in the same way the non-decreasing character of $P_{t+1|t}$.

In Section 4.3 we have seen that initial conditions frequently used both in the case of stationary models and in the case of nonstationary ones are given by $P_0 = \tau I$ with $\tau \gg 0$. It is also possible to obtain Chandrasekhar equations for this situation where $\alpha \leq \min(n,m)$. In order to do this, it is necessary to use the information forms of the Kalman filter. As we have indicated in Section 4.3 this requires the invertibility of matrix Φ, a condition that can fail to be fulfilled in the state-space formulation of econometric models.

Note that the formulation of the dynamic econometric model with measurement errors in the variables, given by equations (3.25)-(3.26), satisfies:

$n_w = m =$ the number of endogenous and exogenous variables in the econometric model.

Thus, if we use either of the three types of initial conditions suggested, the number of equations implied in the use of (A.21)-(A.24) will vary with $2mn$ instead of $n^2/2$, as occurs when the Riccati matrix equation is used.

Finally, we shall see how expressions (A.21) to (A.24) for the suggested algorithm are obtained. The fundamental result to be used is the following, see Morf et al. (1974).

If we define $\delta P_t = P_{t+1|t} - P_{t|t-1}$, it can be shown that

$$\delta P_{t+1} = \left[\Phi - K_{t+1}H\right]\left[\delta P_t + \delta P_t H'(HP_{t|t-1}H' + CRC')^{-1}H\delta P_t\right]\left[\Phi - K_{t+1}H\right]' \quad (A.42)$$

$$= \left[\Phi - K_t H\right]\left[\delta P_t - \delta P_t H'(HP_{t+1|t}H' + CRC')^{-1}H\delta P_t\right]\left[\Phi - K_t H\right]' \quad (A.43)$$

The proof of this result follows. From (A.9) we can write

$$\delta P_{t+1} = \Phi \delta P_t \Phi' - K_{t+1}\left[HP_{t+1|t}H' + CRC'\right]K'_{t+1}$$

$$+ K_t\left[HP_{t|t-1}H' + CRC'\right]K'_t \quad (A.44)$$

From (A.8) we obtain

$$K_{t+1} = \left[\Phi P_{t+1|t}H' + ESC'\right]\left[HP_{t+1|t}H' + CRC'\right]^{-1}$$

$$= \left\{K_t\left[HP_{t|t-1}H' + CRC'\right] + \Phi\delta P_t H'\right\}\left[HP_{t+1|t}H' + CRC'\right]^{-1} \quad (A.45)$$

$$= K_t + \left[\Phi - K_t H\right]\delta P_t H'\left[HP_{t+1|t}H' + CRC'\right]^{-1} \quad (A.46)$$

If we substitute the expression for K_{t+1} given by (A.46) in (A.44), we deduce (A.43). A similar procedure can be used to obtain (A.42).

From (A.42) it follows immediately that $\text{rank}(\delta P_{t+1}) \leq \text{rank}(\delta P_t)$, and it is possible to write $\delta P_t = Y_t X_t Y'_t$, where X_t is a square symmetric matrix whose dimension is given by the rank of δP_0. Bearing the foregoing in mind, we can proceed to obtain expressions (A.21) to

(A.24). (A.21) follows immediately from (A.18) with the definition $\delta P_t = Y_t X_t Y_t'$, and (A.22) is obtained from (A.45). Lastly, (A.23) and (A.24) are obtained by factorizing (A.42).

The alternative expressions (A.23') and (A.24') are obtained by factorizing (A.43) instead of (A.42).

APPENDIX B

CALCULATION OF THE GRADIENT

If $L(\theta)$ is the likelihood function, we want to calculate the gradient of the negative log-likelihood function, $\ell(\theta)$, such that:

$$\ell(\theta) = -\log L(\theta) = \sum_{t=1}^{N} \left[\frac{m}{2}\log(2\pi) + \frac{1}{2}\log|B_t| + \frac{1}{2}\tilde{z}_t' B_t^{-1} \tilde{z}_t \right] \quad (B.1)$$

If θ_i is the i-th element in the vector θ, our aim will be to calculate the general term $\dfrac{\partial \ell(\theta)}{\partial \theta_i}$.

Recall the following matrix derivate results:

$$\frac{\partial \log|A(\alpha)|}{\partial \alpha} = \mathrm{tr}\left[A(\alpha)^{-1} \frac{\partial A(\alpha)}{\partial \alpha} \right] \quad (B.2)$$

$$\frac{\partial A(\alpha)^{-1}}{\partial \alpha} = - A(\alpha)^{-1} \frac{\partial A(\alpha)}{\partial \alpha} A(\alpha)^{-1} \quad (B.3)$$

Differentiating (B.1) we obtain:

$$\frac{\partial \ell(\theta)}{\partial \theta_i} = \sum_{t=1}^{N} \left[\frac{1}{2} \mathrm{tr}\left[B_t^{-1} \frac{\partial B_t}{\partial \theta_i} \right] \right.$$

$$\left. + \tilde{z}_t' B_t^{-1} \frac{\partial \tilde{z}_t}{\partial \theta_i} - \frac{1}{2} \tilde{z}_t' B_t^{-1} \frac{\partial B_t}{\partial \theta_i} B_t^{-1} \tilde{z}_t \right] \quad (B.4)$$

It will be necessary to calculate $\dfrac{\partial \tilde{z}_t}{\partial \theta_i}$ and $\dfrac{\partial B_t}{\partial \theta_i}$, and to do this we will proceed as follows.

From (4.8) we get:

$$\frac{\partial \tilde{z}_t}{\partial \theta_i} = -\frac{\partial H}{\partial \theta_i}\hat{x}_{t|t-1} - H\frac{\partial \hat{x}_{t|t-1}}{\partial \theta_i} - \frac{\partial D}{\partial \theta_i}u_t \qquad (B.5)$$

If we substitute (4.8) into (4.9), we will get:

$$\hat{x}_{t+1|t} = \bar{\Phi}_t \hat{x}_{t|t-1} + \bar{\Gamma}_t u_t + K_t z_t \qquad (B.6)$$

where

$$\bar{\Phi}_t = \Phi - K_t H, \qquad (B.7)$$

$$\bar{\Gamma}_t = \Gamma - K_t D \qquad (B.8)$$

Differentiating (B.6) we obtain:

$$\frac{\partial \hat{x}_{t+1|t}}{\partial \theta_i} = \frac{\partial \bar{\Phi}_t}{\partial \theta_i}\hat{x}_{t|t-1} + \bar{\Phi}_t \frac{\partial \hat{x}_{t|t-1}}{\partial \theta_i} + \frac{\partial \bar{\Gamma}_t}{\partial \theta_i}u_t + \frac{\partial K_t}{\partial \theta_i}z_t \qquad (B.9)$$

The expression $\frac{\partial K_t}{\partial \theta_i}$ is obtained from (4.10):

$$\frac{\partial K_t}{\partial \theta_i} = \left[\frac{\partial \Phi}{\partial \theta_i}P_{t|t-1}H' + \Phi\frac{\partial P_{t|t-1}}{\partial \theta_i}H' + \Phi P_{t|t-1}\frac{\partial H'}{\partial \theta_i}\right.$$

$$+ \frac{\partial E}{\partial \theta_i}SC' + E\frac{\partial S}{\partial \theta_i}C' + ES\frac{\partial C'}{\partial \theta_i}\Bigg] B_t^{-1}$$

$$- \left[\Phi P_{t|t-1}H' + ESC'\right]B_t^{-1}\frac{\partial B_t}{\partial \theta_i}B_t^{-1} \qquad (B.10)$$

As for $\frac{\partial B_t}{\partial \theta_i}$, differentiating (4.12) we obtain:

$$\frac{\partial B_t}{\partial \theta_i} = \frac{\partial H}{\partial \theta_i} P_{t|t-1} H' + H \frac{\partial P_{t|t-1}}{\partial \theta_i} H' + H P_{t|t-1} \frac{\partial H'}{\partial \theta_i}$$

$$+ \frac{\partial C}{\partial \theta_i} RC' + C \frac{\partial R}{\partial \theta_i} C' + CR \frac{\partial C'}{\partial \theta_i} \qquad (B.11)$$

Finally, it will be necessary to calculate the expression for $\frac{\partial P_{t+1|t}}{\partial \theta_i}$. From (4.11) we obtain:

$$\frac{\partial P_{t+1|t}}{\partial \theta_i} = \frac{\partial \Phi}{\partial \theta_i} P_{t|t-1} \Phi' + \Phi \frac{\partial P_{t|t-1}}{\partial \theta_i} \Phi' + \Phi P_{t|t-1} \frac{\partial \Phi'}{\partial \theta_i}$$

$$+ \frac{\partial (EQE')}{\partial \theta_i} - \frac{\partial K_t}{\partial \theta_i} B_t K_t' - K_t \frac{\partial B_t}{\partial \theta_i} K_t' - K_t B_t \frac{\partial K_t'}{\partial \theta_i} \qquad (B.12)$$

If in (B.12) we substitute for $\frac{\partial K_t}{\partial \theta_i}$ its expression (B.10), we can simplify the resulting expression, and if we then substitute for $\frac{\partial B_t}{\partial \theta_i}$ the value given in (B.11), after grouping terms, we get:

$$\frac{\partial P_{t+1|t}}{\partial \theta_i} = \bar{\Phi}_t \frac{\partial P_{t|t-1}}{\partial \theta_i} \bar{\Phi}_t' + A_{it} + A_{it}' \qquad (B.13)$$

where:

$$A_{it} = \frac{\partial \Phi}{\partial \theta_i} P_{t|t-1} \bar{\Phi}_t' - K_t \frac{\partial H}{\partial \theta_i} P_{t|t-1} \bar{\Phi}_t'$$

$$- \frac{\partial E}{\partial \theta_i} SC'K_t' - E \frac{\partial S}{\partial \theta_i} C'K_t' - ES \frac{\partial C'}{\partial \theta_i} K_t'$$

$$+ \frac{1}{2} \frac{\partial (EQE')}{\partial \theta_i} + \frac{1}{2} K_t \frac{\partial (CRC')}{\partial \theta_i} K_t' \qquad (B.14)$$

APPENDIX C

CALCULATION OF THE HESSIAN

The general term, J_{ij}, for the Hessian is given by:

$$J_{ij} = \frac{\partial^2 \ell(\theta)}{\partial \theta_i \partial \theta_j} \tag{C.1}$$

Differentiating (4.16) we get:

$$\begin{aligned}
J_{ij} = \sum_{t=1}^{N} \Bigg\{ & -\frac{1}{2} \operatorname{tr}\left[B_t^{-1} \frac{\partial B_t}{\partial \theta_j} B_t^{-1} \frac{\partial B_t}{\partial \theta_i} \right] + \frac{1}{2} \operatorname{tr}\left[B_t^{-1} \frac{\partial^2 B_t}{\partial \theta_i \partial \theta_j} \right] \\
& + \frac{\partial \tilde{z}_t'}{\partial \theta_j} B_t^{-1} \frac{\partial \tilde{z}_t}{\partial \theta_i} - \tilde{z}_t' B_t^{-1} \frac{\partial B_t}{\partial \theta_j} B_t^{-1} \frac{\partial \tilde{z}_t}{\partial \theta_i} + \tilde{z}_t' B_t^{-1} \frac{\partial^2 \tilde{z}_t}{\partial \theta_i \partial \theta_j} \\
& - \tilde{z}_t' B_t^{-1} \frac{\partial B_t}{\partial \theta_i} B_t^{-1} \frac{\partial \tilde{z}_t}{\partial \theta_j} + \tilde{z}_t' B_t^{-1} \frac{\partial B_t}{\partial \theta_i} B_t^{-1} \frac{\partial B_t}{\partial \theta_j} B_t^{-1} \tilde{z}_t \\
& - \frac{1}{2} \tilde{z}_t' B_t^{-1} \frac{\partial^2 B_t}{\partial \theta_i \partial \theta_j} B_t^{-1} \tilde{z}_t \Bigg\} \tag{C.2}
\end{aligned}$$

The expressions for $\dfrac{\partial^2 \tilde{z}_t}{\partial \theta_i \partial \theta_j}$ and $\dfrac{\partial^2 B_t}{\partial \theta_i \partial \theta_j}$ are obtained directly by differentiating equations (4.17), (4.18), (4.20), (4.21) and (4.22). While the results are analytically complex, their programming is simple. However the computational requirements are considerable, approximately equivalent to the solution of $\left[n(n+1)p^2\right]/2$ equations.

If the terms in second derivatives and the penultimate term in expression (C.2) are negligible, the following expression would be a valid approximation:

$$J_{ij} \approx \sum_{t=1}^{N} \left[-\frac{1}{2} \text{tr}\left[B_t^{-1} \frac{\partial B_t}{\partial \theta_j} B_t^{-1} \frac{\partial B_t}{\partial \theta_i} \right] + \frac{\partial \tilde{z}_t'}{\partial \theta_j} B_t^{-1} \frac{\partial \tilde{z}_t}{\partial \theta_i} \right.$$

$$\left. - \tilde{z}_t' B_t^{-1} \frac{\partial B_t}{\partial \theta_j} B_t^{-1} \frac{\partial \tilde{z}_t}{\partial \theta_i} - \tilde{z}_t' B_t^{-1} \frac{\partial B_t}{\partial \theta_i} B_t^{-1} \frac{\partial \tilde{z}_t}{\partial \theta_j} \right] \quad (C.3)$$

The calculation of this expression uses information obtained in the evaluation of the gradient, and so it does not involve any additional computational requirement. This approximation could be valid in practice depending upon the parametrization of the problem.

APPENDIX D

CALCULATION OF THE INFORMATION MATRIX

The general term for the information matrix is:

$$\left[M(\theta)\right]_{ij} = E\left[\left.\frac{\partial^2 \ell(\theta)}{\partial \theta_i \partial \theta_j}\right|_\theta\right] \tag{D.1}$$

We can write the expression (C.2) obtained in Appendix C for $\frac{\partial^2 \ell(\theta)}{\partial \theta_i \partial \theta_j}$ in the form:

$$\begin{aligned}
J_{ij} = \sum_{t=1}^{N} &\left\{ -\frac{1}{2} \text{tr}\left[B_t^{-1} \frac{\partial B_t}{\partial \theta_i} B_t^{-1} \frac{\partial B_t}{\partial \theta_j}\right] + \frac{1}{2} \text{tr}\left[B_t^{-1} \frac{\partial^2 B_t}{\partial \theta_i \partial \theta_j}\right] \right. \\
&+ \text{tr}\left[B_t^{-1} \frac{\partial \tilde{z}_t}{\partial \theta_i} \frac{\partial \tilde{z}_t'}{\partial \theta_j}\right] - \text{tr}\left[B_t^{-1} \frac{\partial B_t}{\partial \theta_j} B_t^{-1} \frac{\partial \tilde{z}_t}{\partial \theta_i} \tilde{z}_t'\right] \\
&+ \text{tr}\left[B_t^{-1} \frac{\partial^2 \tilde{z}_t}{\partial \theta_i \partial \theta_j} \tilde{z}_t'\right] - \text{tr}\left[B_t^{-1} \frac{\partial B_t}{\partial \theta_i} B_t^{-1} \frac{\partial \tilde{z}_t}{\partial \theta_j} \tilde{z}_t'\right] \\
&\left. + \text{tr}\left[B_t^{-1} \frac{\partial B_t}{\partial \theta_i} B_t^{-1} \frac{\partial B_t}{\partial \theta_j} B_t^{-1} \tilde{z}_t \tilde{z}_t'\right] - \frac{1}{2} \text{tr}\left[B_t^{-1} \frac{\partial^2 B_t}{\partial \theta_i \partial \theta_j} B_t^{-1} \tilde{z}_t \tilde{z}_t'\right] \right\}
\end{aligned} \tag{D.2}$$

If we bear in mind that $E\left[\tilde{z}_t\right] = 0$ and that, according to equation (4.17), $\frac{\partial \tilde{z}_t}{\partial \theta_i}$ is not a function of $\tilde{z}_t$, then:

$$E\left[\frac{\partial \tilde{z}_t}{\partial \theta_i} \tilde{z}_t'\right] = 0, \quad E\left[\frac{\partial^2 \tilde{z}_t}{\partial \theta_i \partial \theta_j} \tilde{z}_t'\right] = 0$$

If we take expected values in (D.2), we get:

$$[M(\theta)]_{ij} = \sum_{t=1}^{N} \left\{ \frac{1}{2} \text{tr}\left[B_t^{-1} \frac{\partial B_t}{\partial \theta_i} B_t^{-1} \frac{\partial B_t}{\partial \theta_j}\right] + \text{tr}\left[B_t^{-1} E\left[\frac{\partial \tilde{z}_t}{\partial \theta_i} \frac{\partial \tilde{z}_t'}{\partial \theta_j}\right]\right] \right\} \quad \text{(D.3)}$$

This expression coincides with that used in Watson and Engle (1983), and obtained in Engle and Watson (1981).

We will now have to calculate the expression corresponding to $E\left[\frac{\partial \tilde{z}_t}{\partial \theta_i} \frac{\partial \tilde{z}_t'}{\partial \theta_j}\right]$. To do this, we first observe that:

$$E\left[\frac{\partial \tilde{z}_t}{\partial \theta_i} \frac{\partial \tilde{z}_t'}{\partial \theta_j}\right] = B_t^{ij} + \frac{\overline{\partial \tilde{z}_t}}{\partial \theta_i} \frac{\overline{\partial \tilde{z}_t'}}{\partial \theta_j} \quad \text{(D.4)}$$

where $\frac{\overline{\partial \tilde{z}_t}}{\partial \theta_i}$ is the mean of $\frac{\partial \tilde{z}_t}{\partial \theta_i}$ and B_t^{ij} is the covariance matrix of $\frac{\partial \tilde{z}_t}{\partial \theta_i}$ and $\frac{\partial \tilde{z}_t}{\partial \theta_j}$. Thus:

$$[M(\theta)]_{ij} = \sum_{t=1}^{N} \left\{ \frac{1}{2} \text{tr}\left[B_t^{-1} \frac{\partial B_t}{\partial \theta_i} B_t^{-1} \frac{\partial B_t}{\partial \theta_j}\right] + \text{tr}\left[B_t^{-1} B_t^{ij}\right] \right.$$
$$\left. + \text{tr}\left[B_t^{-1} \frac{\overline{\partial \tilde{z}_t}}{\partial \theta_i} \frac{\overline{\partial \tilde{z}_t'}}{\partial \theta_j}\right] \right\} \quad \text{(D.5)}$$

The problem is then reduced to the calculation of $\frac{\overline{\partial \tilde{z}_t}}{\partial \theta_i}$, $\frac{\overline{\partial \tilde{z}_t}}{\partial \theta_j}$ and of B_t^{ij}.

If we note that expression (4.18) is equivalent to:

$$\frac{\partial \hat{x}_{t+1|t}}{\partial \theta_i} = \bar{\Phi}_t \frac{\partial \hat{x}_{t|t-1}}{\partial \theta_i} + \left[\frac{\partial \Phi}{\partial \theta_i} - K_t \frac{\partial H}{\partial \theta_i} \right] \hat{x}_{t|t-1}$$

$$+ \left[\frac{\partial \Gamma}{\partial \theta_i} - K_t \frac{\partial D}{\partial \theta_i} \right] u_t + \frac{\partial K_t}{\partial \theta_i} \tilde{z}_t$$

taking equations (4.9) and (4.17) into account, we can write the following linear system of dimension 3n:

$$x_{t+1}^c = \Phi_t^c x_t^c + \Gamma_t^c u_t + K_t^c \tilde{z}_t \qquad (D.6)$$

$$z_t^c = H^c x_t^c + D^c u_t \qquad (D.7)$$

where

$$x_t^c = \left[\hat{x}_{t|t-1}' \quad \frac{\partial \hat{x}_{t|t-1}'}{\partial \theta_i} \quad \frac{\partial \hat{x}_{t|t-1}'}{\partial \theta_j} \right]'$$

$$z_t^c = \left[\frac{\partial \tilde{z}_t'}{\partial \theta_i} \quad \frac{\partial \tilde{z}_t'}{\partial \theta_j} \right]'$$

and such that:

$$\Phi_t^c = \begin{bmatrix} \Phi & 0 & 0 \\ \frac{\partial \Phi}{\partial \theta_i} - K_t \frac{\partial H}{\partial \theta_i} & \bar{\Phi}_t & 0 \\ \frac{\partial \Phi}{\partial \theta_j} - K_t \frac{\partial H}{\partial \theta_j} & 0 & \bar{\Phi}_t \end{bmatrix}, \quad \Gamma_t^c = \begin{bmatrix} \Gamma \\ \frac{\partial \Gamma}{\partial \theta_i} - K_t \frac{\partial D}{\partial \theta_i} \\ \frac{\partial \Gamma}{\partial \theta_j} - K_t \frac{\partial D}{\partial \theta_j} \end{bmatrix}$$

$$K_t^C = \begin{bmatrix} K_t \\ \dfrac{\partial K_t}{\partial \theta_i} \\ \dfrac{\partial K_t}{\partial \theta_j} \end{bmatrix}, \quad H^C = \begin{bmatrix} -\dfrac{\partial H}{\partial \theta_i} & -H & 0 \\ -\dfrac{\partial H}{\partial \theta_j} & 0 & -H \end{bmatrix}, \quad D^C = \begin{bmatrix} -\dfrac{\partial D}{\partial \theta_i} \\ -\dfrac{\partial D}{\partial \theta_j} \end{bmatrix}$$

If we define

$$\bar{x}_t^C = E\left[x_t^C\right], \quad \bar{z}_t^C = E\left[z_t^C\right]$$

and

$$P_t^C = E\left\{\left[x_t^C - \bar{x}_t^C\right]\left[x_t^C - \bar{x}_t^C\right]'\right\}, \quad B_t^C = E\left\{\left[z_t^C - \bar{z}_t^C\right]\left[z_t^C - \bar{z}_t^C\right]'\right\}$$

The mean and covariance equations of the system are propagated as follows:

$$\bar{x}_{t+1}^C = \Phi_t^C \bar{x}_t^C + \Gamma_t^C u_t \tag{D.8}$$

$$\bar{z}_t^C = H^C \bar{x}_t^C + D^C u_t \tag{D.9}$$

$$P_{t+1}^C = \Phi_t^C P_t^C (\Phi_t^C)' + K_t^C B_t (K_t^C)' \tag{D.10}$$

$$B_t^C = H^C P_t^C (H^C)' \tag{D.11}$$

Thus we get the values of $\dfrac{\partial \tilde{z}_t}{\partial \theta_i}$ and $\dfrac{\partial \tilde{z}_t}{\partial \theta_j}$ from equations (D.8) and (D.9). The value of B_t^{ij} is given by (D.10) and (D.11) given that:

$$B_t^C = \begin{bmatrix} B_t^{ii} & B_t^{ij} \\ B_t^{ji} & B_t^{jj} \end{bmatrix}$$

The initial conditions for these equations are:

$$x_0^c = \begin{bmatrix} \bar{x}_0' & 0 & 0 \end{bmatrix}'$$

$$P_0^c = \begin{bmatrix} P_0 & 0 & 0 \\ 0 & 0 & 0 \\ 0 & 0 & 0 \end{bmatrix}$$

As indicated by (4.32), when there is no constant term and the system is stationary $\bar{x}_0 = 0$ and P_0 will be given by the solution of the algebraic Lyapunov equation (4.33). In nonstationary situations, the value of $\bar{x}_0$ is given by (E.13) and $P_0 = 0$.

APPENDIX E

ESTIMATION OF THE INITIAL CONDITIONS

As we have pointed out, the algorithm for the estimation of the model requires the initial conditions of the system set out in (4.1) and (4.2). If we take the initial state x_0 as an unknown but constant parameter, we can use $P_0 = 0$. In this appendix we will obtain a maximum likelihood estimator $\hat{x}_0$, for x_0. We will also show that $\hat{x}_0$ is the result of the solution of a system of linear equations.

The vector θ of p parameters to estimate will be:

$$\theta = \begin{bmatrix} \theta_1' & x_0' \end{bmatrix}' \tag{E.1}$$

where θ_1 is the vector of (p - n) unknown parameters of the matrices of the system. Recall that the function to be minimized is given by:

$$\ell(\theta) = \sum_{t=1}^{N} \left[\frac{m}{2} \log(2\pi) + \frac{1}{2} \log|B_t| + \frac{1}{2} \tilde{z}_t' B_t^{-1} \tilde{z}_t \right] \tag{E.2}$$

where:

$$\tilde{z}_t = z_t - H\hat{x}_{t|t-1} - Du_t \tag{E.3}$$

is the innovations process and B_t its covariance matrix. Substituting (4.8) into (4.9) we get:

$$\hat{x}_{t+1|t} = \bar{\Phi}_t \hat{x}_{t|t-1} + \bar{\Gamma}_t u_t + K_t z_t \tag{E.4}$$

where:

$$\bar{\Phi}_t = \Phi - K_t H, \quad \bar{\Gamma}_t = \Gamma - K_t D \tag{E.5}$$

If instead of the initial conditions given by x_0 and $P_0 = 0$ we use x_0^* and $P_0 = 0$, we will get different expressions for the estimates of the state vector and for the innovation process, which we shall represent by $\hat{x}_{t+1|t}^*$ and $\tilde{z}_t^*$. It is then easy to show that:

$$\hat{x}_{t+1|t} = \hat{x}_{t+1|t}^* + \bar{\bar{\Phi}}_{t+1}(x_0 - x_0^*) \tag{E.6}$$

where:

$$\bar{\bar{\Phi}}_{t+1} = \bar{\Phi}_{t+1}\bar{\bar{\Phi}}_t \tag{E.7}$$

or, by virtue of (E.5):

$$\bar{\bar{\Phi}}_{t+1} = (\Phi - K_{t+1}H)\bar{\bar{\Phi}}_t \quad \text{with } \bar{\bar{\Phi}}_0 = I \tag{E.8}$$

Expression (E.6) is obtained observing that the matrices Φ, Γ, H, D and K_t depend only on θ_1 and not on x_0. Thus when we change only the initial conditions of x_0 to x_0^*, these matrices remain unchanged.

From (E.3) and (E.6) we get:

$$\tilde{z}_t = \tilde{z}_t^* - H\bar{\bar{\Phi}}_t(x_0 - x_0^*) \tag{E.9}$$

As B_t is not a function of the initial conditions either, the expression for $\ell(\theta)$ given by (E.2), and in view of (E.9), is a quadratic expression in x_0. Thus the following is a linear function in x_0:

$$\frac{\partial \ell(\theta_1, x_0)}{\partial x_0} = 0 \tag{E.10}$$

The solution to equation (E.10) will give an analytical expression for the maximum likelihood estimate of x_0. In fact, from (E.2) we get:

$$\frac{\partial \ell(\theta_1, x_0)}{\partial x_0} = \sum_{t=1}^{N} \tilde{z}_t' B_t^{-1} \frac{\partial \tilde{z}_t}{\partial x_0} = 0 \tag{E.11}$$

The following results from (E.9):

$$\frac{\partial \tilde{z}_t}{\partial x_0} = -\bar{\bar{H\Phi}}_t \tag{E.12}$$

Substituting (E.9) and (E.12) into (E.11), we will get the estimator of the initial state vector given by:

$$\hat{x}_0 = \left[\sum_{t=1}^{N} \bar{\bar{\Phi}}_t' H' B_t^{-1} \bar{\bar{H\Phi}}_t\right]^{-1} \left[\sum_{t=1}^{N} \bar{\bar{\Phi}}_t' H' B_t^{-1} (\tilde{z}_t^* + \bar{\bar{H\Phi}}_t x_0^*)\right] \tag{E.13}$$

The previous expression corresponds to the minimum for $\ell(\theta_1, x_0)$, because

$$\frac{\partial^2 \ell(\theta_1, x_0)}{\partial x_0 \, \partial x_0'} = \sum_{t=1}^{N} \bar{\bar{\Phi}}_t' H' B_t^{-1} \bar{\bar{H\Phi}}_t \tag{E.14}$$

is always positive definite since $B_t = HP_{t|t-1}H' + CRC' > 0$, as $P_{t|t-1}$ is a covariance matrix and therefore nonnegative and $R > 0$.

Thus we can see how (E.13) gives an analytical expression for the likelihood estimation of x_0 given the other parameters of the model.

Note that the expression for $\hat{x}_0$ assumes that the matrix $\sum_{t=1}^{N} \bar{\bar{\Phi}}_t' H' B_t^{-1} \bar{\bar{H\Phi}}_t$ has an inverse. However, the rank of this matrix is in general only equal or greater than m, since $B_t > 0$ and rank(H) = m. On the other hand, a solution to (E.11) always exists, since it can easily be shown that:

$$\sum_{t=1}^{N} \bar{\bar{\Phi}}_t' H' B_t^{-1} (\tilde{z}_t^* + \bar{\bar{H\Phi}}_t x_0^*) \in \text{range}\left[\sum_{t=1}^{N} \bar{\bar{\Phi}}_t' H' B_t^{-1} \bar{\bar{H\Phi}}_t\right]$$

Thus, when m < n, the solution is not necessarily unique. In such a case, the x_0 vector is not identifiable and to get a unique estimate it would be necessary to incorporate additional information about (n - m) of its components. Nevertheless, given that the parameters whose estimation we are interested in correspond to the vector θ_1, an acceptable solution when m < n is to use a generalized inverse in (E.13). As we have stated repeatedly, the situation where m < n is to be found frequently in econometric models.

Indeed, the solution here avoids the problem involved in estimating simultaneously all the parameters of the vector defined in (E.1), since it is sufficient to use the algorithm of Chapter 4 to minimize $\ell(\theta)$ with respect to θ_1. The minimization with respect to x_0 can be solved analytically, as given by (E.13), and is not coupled with the previous one.

The basic idea in this analysis is to use expressions (E.6) and (E.9) to obtain the corresponding magnitudes for the Kalman filter, with respect to a change in the initial conditions. In this context the results given here generalize those given by Rosenberg (1973) and Harvey and McKenzie (1984), and amount to a specific case of the more general sensitivity analysis presented by Terceiro (1975).

APPENDIX F

SOLUTION OF THE LYAPUNOV AND RICCATI EQUATIONS

F.1. Lyapunov Equation

The Lyapunov equation is of the following form:

$$P_{t+1} = \Phi P_t \Phi' + \bar{Q} \tag{F.1}$$

where the P_t, Φ and $\bar{Q}$ matrices are (n x n) and P_t and $\bar{Q}$ are symmetric. The corresponding algebraic equation whose solution is of interest will be:

$$P = \Phi P \Phi' + \bar{Q} \tag{F.2}$$

A first procedure for the calculation of P in (F.2) is to use (F.1), iterating from an initial value P_0, so that

$$P = \lim_{t \to \infty} P_t \tag{F.3}$$

Note that the analytical solution to the equation (F.1) is given by

$$P_t = \Phi^t P_0 (\Phi')^t + \sum_{i=0}^{t-1} \Phi^i \bar{Q} (\Phi')^i \tag{F.4}$$

Another procedure for the solution of (F.2) consists in previously vectorizing the P and $\bar{Q}$ matrices in such a way as to enable us to write (F.2) as follows:

$$\text{vec}(P) = (\Phi \otimes \Phi)\,\text{vec}(P) + \text{vec}(\bar{Q}) \tag{F.5}$$

where the vec(.) operator consists in transforming the matrix in a vector whose components are the successive columns of the matrix. The symbol ⊗ represents the Kronecker product. From (F.5) we can obtain the following direct solution:

$$\text{vec}(P) = \left[I - \Phi \otimes \Phi\right]^{-1} \text{vec}(\bar{Q}) \qquad (F.6)$$

Note that the eigenvalues of the $I - \Phi \otimes \Phi$ matrix are $1 - \lambda_i \lambda_j$, where λ_i are the eigenvalues of Φ. Thus a necessary and sufficient condition for (F.2) to have a unique solution is that $\lambda_i \lambda_j \neq 1$. Obviously the stability of Φ is enough to insure this.

The solution given by (F.6) is not computationally efficient in situations where n is large, given that the number of multiplications required varies with n^6.

The third procedure which we put forward for the solution of (F.2) does not have such heavy computational requirements because the number of multiplications varies with n^3. This algorithm, due to Barraud (1977), has two steps. First, the algebraic Lyapunov equation (F.2) is transformed into the equivalent equation

$$P^* = \Phi^* P^* (\Phi^*)' + \bar{Q}^* \qquad (F.7)$$

where Φ^* is the upper real Schur form. Second (F.7) is solved by back substitution.

The (F.7) equation is obtained by transforming the Φ matrix to the upper Hessenberg form with the orthogonal matrix U, so that

$$U\Phi U' = \Phi^+ \qquad (F.8)$$

This can be done with a Householder transformation. Next, the orthogonal matrix V is obtained, by means of the QR algorithm, so that

$$\Phi^* = V\Phi^+ V' = VU\Phi U'V' = W\Phi W' \qquad (F.9)$$

where W is an orthogonal matrix.

If we define

$$P^* = WPW'$$

$$\bar{Q}^* = W\bar{Q}W'$$

we immediately see that equations (F.2) and (F.7) are equivalent. Note that all these transformations can be obtained using standard software, see Rice (1983).

Given that Φ^* is in the upper Schur form it can be partioned into (s x s) blocks with a maximum dimension of (2 x 2). We shall represent these blocks by Φ^*_{ij} and they are such that $\Phi^*_{ij} = 0$ for $i > j$. That is to say, the structure of Φ^* will be

$$\Phi^* = \begin{bmatrix} \Phi^*_{11} & \Phi^*_{12} & \cdots & \Phi^*_{1s} \\ 0 & \Phi^*_{22} & \cdots & \Phi^*_{2s} \\ \cdot & \cdot & \cdots & \cdot \\ \cdot & \cdot & \cdots & \cdot \\ \cdot & \cdot & \cdots & \cdot \\ 0 & 0 & \cdots & \Phi^*_{ss} \end{bmatrix}$$

Thus equation (F.7) can be written as follows:

$$P^*_{kh} = \sum_{i=1}^{k} \Phi^*_{ik} \left[\sum_{j=1}^{h} P^*_{ij}(\Phi^*)'_{kh} \right] + \bar{Q}^*_{kh} \qquad (F.10)$$

$$h = 1,\ldots,s$$
$$k = h,\ldots,s$$

These equations should be solved successively for $P^*_{11},\ldots,P^*_{s1}$, P^*_{22}, $\ldots$, $P^*_{s2},\ldots,P^*_{ss}$. Each block P^*_{kh} implies the solution of a first or second order Lyapunov equation. In effect, if we consider the calculation of P^*_{kh} by developing (F.10) we obtain

$$P^*_{kh} = \sum_{i=1}^{k-1} \Phi^*_{ik} \left[\sum_{j=1}^{h} P^*_{ij}(\Phi^*_{kh})' \right] + \Phi^*_{kk} \sum_{j=1}^{h-1} P^*_{kj}(\Phi^*)'_{jh}$$

$$+ \Phi^*_{kk} P^*_{kh}(\Phi^*_{hh})' + \bar{Q}^*_{kh} \qquad (F.11)$$

Note that the first two components of the second term of the previous equation are known (or zero, when $k = h = 1$). Thus (F.11) can be written as follows:

$$P^*_{kh} = \Phi^*_{kk} P^*_{kh} (\Phi^*_{hh})' + \bar{\bar{Q}}^*_{kh} \qquad (F.12)$$

where

$$\bar{\bar{Q}}^*_{kh} = \bar{Q}^*_{kh} + \sum_{i=1}^{k-1} \Phi^*_{ik} \left[\sum_{j=1}^{h} P^*_{ij} (\Phi^*_{kh})' \right] + \Phi^*_{kk} \sum_{j=1}^{h-1} P^*_{kj} (\Phi^*)'_{jh}$$

We see then that the problem is reduced to solving equations of the (F.12) type, which are first or second order Lyapunov equations, and which can be non-symmetric.

Finally, once P^* is calculated, the solution to (F.2) is given by

$$P = W'P^*W \qquad (F.13)$$

Note that, in our estimation problem as analyzed in Chapter 4, we require the solution of Lyapunov equations for determining the initial conditions for the Kalman filter in stationary systems given by (4.32)-(4.33). Their solution is also required when the stationary version of the sensitivity equations of the covariance matrix of the Kalman filter given by (4.25) is used in the calculation of the gradient of the likelihood function. The estimation algorithm described requires the solution of one of these equations for each parameter to be estimated, that is p equations. Nevertheless, it should be noted that these equations differ from each other only in the constant term $(A_i + A'_i)$. So only a transformation to the upper Schur form is required, which considerably simplifies the solution of these p equations.

F.2. Riccati Equation

The matrix equations for the Kalman filter given by (4.10), (4.11), and (4.12) allow the following Riccati equation to be written:

$$P_{t+1|t} = \Phi P_{t|t-1}\Phi' - \left[\Phi P_{t|t-1}H' + ESC'\right]\left[HP_{t|t-1}H' + CRC'\right]^{-1}$$

$$\left[\Phi P_{t|t-1}H' + ESC'\right]' + EQE' \tag{F.14}$$

Thus, the corresponding algebraic matrix Riccati equation will be

$$P = \Phi P \Phi' - \left[\Phi PH' + ESC'\right]\left[HPH' + CRC'\right]^{-1}\left[\Phi PH' + ESC'\right]'$$

$$+ EQE' \tag{F.15}$$

where we represent the stationary solution of (F.14) by P.

The conditions under which P can exist are studied in detail in Chan et al. (1984), and we shall not refer to them here. But it is interesting to note that under certain conditions P exists even when the Φ matrix has eigenvalues on the unit circle.

The solution to (F.15) can be obtained by applying iterative procedures to (F.14), on the basis of an initial positive semidefinite matrix P_0, so that

$$P = \lim_{t \to \infty} P_{t+1|t} \tag{F.16}$$

Instead of iterating by using expression (F.14), it is convenient to use the equivalent expression:

$$P_{t+1|t} = \bar{\Phi}_t P_{t|t-1}\bar{\Phi}'_t + \left[E - K_t C\right]\begin{bmatrix} Q & S \\ S' & R \end{bmatrix}\begin{bmatrix} E' \\ -C'K'_t \end{bmatrix} \tag{F.17}$$

As pointed out in Appendix A, equation (F.17) is more stable numerically than (F.14).

The second of the procedures recommended for solving (F.15) is a generalization of the Hewer (1971) method, which is based on the following result:

If P_t, $t = 0,1,\ldots,$ is the only positive semidefinite solution of the following equation

$$P_t = \bar{\Phi}_t P_t \bar{\Phi}_t' + EQE' + K_t CRC'K_t' - K_t CS'E' - ESC'K_t' \qquad (F.18)$$

where

$$K_t = \left[\Phi P_{t-1}H' + ESC'\right]\left[HP_{t-1}H' + CRC'\right]^{-1} \qquad (F.19)$$

and

$$\bar{\Phi}_t = \Phi - K_t H \qquad (F.20)$$

and if we select K_0 in such a way that the matrix $\bar{\Phi}_0 = \Phi - K_0 H$ is stable, it can be shown that

i) $0 \leq P \leq P_{t+1} \leq P_t < \ldots < P_0$

ii) $\lim_{t \to \infty} P_t = P$

iii) in the vicinity of P, $\|P_{t+1} - P\| \leq \tau \|P_t - P\|^2$ where τ is a finite constant and $\|P_t - P\|$ is a consistent matrix norm.

The demonstration of this result can be developed following Hewer (1971), and for this reason we do not include it here.

Note that equation (F.18) is, in its t-th iteration, a Lyapunov equation, since the four last summands of the second term are independent of P_t.

In short, the iterative process proposed for the solution of (F.15) has the following stages:

a) Choose K_0 in such a way that $\bar{\Phi}_0$ is stable (i.e. has eigenvalues whose magnitudes are less than unity).
b) Calculate P_0 from the Lyapunov equation (F.18).
c) Calculate K_1 from (F.19).
d) Return to b) to calculate P_1.

The process will be stopped when $\|K_{t+1} - K_t\| < \eta$, where η is a previously fixed scalar. Note that each iteration requires the solution of a Lyapunov equation.

There are other procedures, see Pappas et al. (1980), for the solution of the Riccati equations which are based on the solution of a generalized eigenvalue problem. The main feature of these new methods is that they do not require inversion of the matrix Φ of the model. Thus, they are directly applicable to the dynamic econometric model formulated in Chapter 2 with singular transition matrices Φ, which very often occur in practice. We do not analyze these procedures here because they imply the calculation of the generalized eigenvalues and eigenvectors of matrices of order 2n. In many cases, this implies a very high computational burden since, as we have pointed out, the formulation of econometric models in state-space tends to be characterized by large values of n.

APPENDIX G

FIXED-INTERVAL SMOOTHING ALGORITHM

We begin with the general state-space formulation given by:

$$x_{t+1} = \Phi x_t + \Gamma u_t + E w_t \tag{G.1}$$

$$z_t = H x_t + D u_t + C v_t \tag{G.2}$$

If one wants to treat problems of missing observations or of contemporaneous or temporal aggregation of observations, in either the endogenous or the exogenous variables, one must include those exogenous variables that are characterized by such situations in the vector z_t. This is done according to the procedure used in Chapter 3 for the treatment of exogenous variables observed with error.

Therefore, we assume that in (G.1)-(G.2) the exogenous variables explicitly considered in the vector u_t are those for which all observations are available and free of error. All of the other exogenous variables are included with the endogenous variables in the vector z_t.

Following the justifications given in Chapter 5, we assume that the observed variables are not the m components of the vector z, but the m^* components of the vector z_t^* defined by:

$$z_t^* = H_t^* z_t \tag{G.3}$$

where in general $H_t^* \neq I_m$.

Under such conditions, the true observation equation, substituting for (G.2), is:

$$z_t^* = \bar{H}_t x_t + \bar{D}_t u_t + \bar{C}_t v_t \tag{G.2'}$$

where

$$\bar{H}_t = H_t^* H \, , \quad \bar{D}_t = H_t^* D \, , \quad \bar{C}_t = H_t^* C$$

Note that the matrices $\bar{H}_t$, $\bar{D}_t$ and $\bar{C}_t$ depend on time in the general case.

If we define:

$$\hat{x}_{i|j} = E\left[x_i \big| (Z^*)^j\right] \, , \quad (Z^*)^j = \left[(z_1^*)', \ldots, (z_j^*)'\right]' \qquad (G.4)$$

$$P_{i|j} = E\left[(x_i - \hat{x}_{i|j})(x_i - \hat{x}_{i|j})' \big| (Z^*)^j\right] \qquad (G.5)$$

we obtain, for $j = n$, the following fixed-interval smoothing algorithm:

$$\hat{x}_{t|n} = \hat{x}_{t|t} + A_t(\hat{x}_{t+1|n} - \hat{x}_{t+1|t}) \qquad (G.6)$$

$$A_t = P_{t|t} \Phi' P_{t+1|t}^+ \qquad (G.7)$$

$$P_{t|n} = P_{t|t} - A_t(P_{t+1|t} - P_{t+1|n})A_t' \qquad (G.8)$$

The superindex (+) denotes the generalized inverse. It is important to use a high-quality and computationally efficient algorithm for the generalized inverse. Expressions (G.6)-(G.8) can be deduced by a procedure analogous to that used in Appendix A for the Kalman filter algorithm, see Kohn and Ansley (1983).

For $i \geq j$ the values of $\hat{x}_{i|j}$ and $P_{i|j}$ used in the smoothing algorithm (G.6)-(G.8) are given by the following Kalman filter:

$$\tilde{z}_t^* = z_t^* - \bar{H}_t \hat{x}_{t|t-1} - \bar{D}_t u_t \qquad (G.9)$$

$$\bar{B}_t = \bar{H}_t P_{t|t-1} \bar{H}_t' + \bar{C}_t R \bar{C}_t' \qquad (G.10)$$

$$\bar{K}_t = P_{t|t-1} \bar{H}_t' \bar{B}_t^{-1} \qquad (G.11)$$

$$\hat{x}_{t|t} = \hat{x}_{t|t-1} + \bar{K}_t \tilde{z}_t^* \tag{G.12}$$

$$P_{t|t} = (I - \bar{K}_t \bar{H}_t) P_{t|t-1} \tag{G.13}$$

$$\hat{x}_{t+1|t} = \Phi \hat{x}_{t|t} + \Gamma u_t + E S \bar{C}_t' \bar{B}_t^{-1} \tilde{z}_t^* \tag{G.14}$$

$$P_{t+1|t} = \Phi P_{t|t} \Phi' + EQE' - ES\bar{C}_t'\bar{B}_t^{-1}\bar{C}_t S'E' - \Phi \bar{K}_t \bar{C}_t S'E' - ES\bar{C}_t'\bar{K}_t \Phi' \tag{G.15}$$

The initial conditions will be:

$$\hat{x}_{0|-1} = \bar{x}_0 \, , \quad P_{0|-1} = P_0 \tag{G.16}$$

Note that equations (G.9)-(G.16) are equivalent to (A.6)-(A.10) obtained in Appendix A, using (G.2') instead of (A.2), and making the updating cycle given by (G.9)-(G.13) and the propagation given by (G.14)-(G.15) explicit.

One may be interested not only in estimating the econometric model under the assumption that the variables z_t^* are observed as in (G.2'), but also in obtaining estimates of the "observations" corresponding to points in time in which we do not have these values or only have them in aggregate form. That is, one wants expressions for:

$$\hat{z}_{t|n} = E\left[z_t \mid (z^*)^n\right] \, , \quad (z^*)^n = \left[(z_1^*)', \ldots, (z_n^*)'\right]' \tag{G.17}$$

$$P_{t|n}^z = E\left[(z_t - \hat{z}_{t|n})(z_t - \hat{z}_{t|n})' \mid (z^*)^n\right] \tag{G.18}$$

It is easy to demonstrate, in a way analogous to that used in Kohn and Ansley (1983), the following result:

$$\hat{z}_{t|n} = H\hat{x}_{t|n} + Du_t + \bar{A}_t(\hat{x}_{i+1|n} - \hat{x}_{i+1|i}) + \bar{\bar{A}}_t \tilde{z}_t^* \tag{G.19}$$

where

$$\bar{A}_t = (CS'E' - CR\bar{C}'_t \bar{\bar{K}}'_t) P^+_{t+1|t} \tag{G.20}$$

$$\bar{\bar{K}}_t = (\Phi P_{t|t-1} \bar{H}' + ES\bar{C}'_t) \bar{B}^+_t \tag{G.21}$$

$$\bar{\bar{A}}_t = CR\bar{C}'_t \bar{B}^+_t \tag{G.22}$$

One can also obtain

$$P^z_{t|n} = H_t P_{t|t} H'_t + CRC' - CR\bar{C}'_t \bar{K}'_t H' - H\bar{K}_t \bar{C}_t RC'$$

$$\quad - (\bar{A}_t + HA_t)(P_{t+1|t} - P_{t+1|n})(\bar{A}_t + HA_t)'$$

$$\quad - \bar{\bar{A}}_t B_t \bar{\bar{A}}'_t \tag{G.23}$$

An alternative procedure for calculating $\hat{z}_{t|n}$ and $P^z_{t|n}$, that does not require the use of (G.19)-(G.23), consists in defining the expanded state vector $\begin{bmatrix} x'_t & z'_{t-1} \end{bmatrix}'$, so that equations (G.1)-(G.3) can be written in the form:

$$\begin{bmatrix} x_{t+1} \\ z_t \end{bmatrix} = \begin{bmatrix} \Phi & 0 \\ H & 0 \end{bmatrix} \begin{bmatrix} x_t \\ z_{t-1} \end{bmatrix} + \begin{bmatrix} \Gamma \\ D \end{bmatrix} u_t + \begin{bmatrix} E & 0 \\ 0 & C \end{bmatrix} \begin{bmatrix} w_t \\ v_t \end{bmatrix} \tag{G.24}$$

$$z^*_t = \begin{bmatrix} \bar{H}_t & 0 \end{bmatrix} \begin{bmatrix} x_t \\ z_{t-1} \end{bmatrix} + \bar{D}_t u_t + \bar{C}_t v_t \tag{G.25}$$

We can then apply the fixed-interval smoothing algorithm given by (G.6)-(G.8) to the system (G.24)-(G.25) thus defined to obtain $\hat{z}_{t|n}$ and $P^z_{t|n}$ directly. Since the dimension of x_t in (G.1) is frequently much higher than that of z_t in (G.2), the increase in the computational burden resulting from use of (G.24)-(G.25) may not be large.

REFERENCES

Aigner, D.J., Hsiao, C., Kapteyn, A. and Wansbech, T. (1984). Latent variable models in econometrics, in Z. Griliches and M.D. Intriligator (editors), Handbook of Econometrics, vol. II. Amsterdam: North-Holland.

Akaike, H. (1974). A new look at the statistical model identification. IEEE Transactions on Automatic Control, vol. AC-19, no. 6, pp. 716-723.

Anderson, B.D.O. and Deistler, M. (1984). Identifiability in dynamic errors in variables models. Journal of Time Series Analysis, vol. 5, no. 1, pp.1-13.

Anderson, B.D.O. and Moore, J.B. (1979). Optimal Filtering. Englewood Cliffs, N.J.: Prentice-Hall.

Ansley, C.F. and Kohn, R. (1983). Exact likelihood of vector autoregressive-moving average process with missing or aggregate data. Biometrika, vol. 70, no. 1, pp. 275-278.

Aoki, M. (1976). Optimal Control and System Theory in Dynamic Analysis. Amsterdam: North-Holland.

Aoki, M. (1987). State Space Modelling of Time Series. New York: Springer Verlag.

Bar-Shalom, Y. (1971). On the asymptotic properties of the maximum-likelihood estimate obtained from dependent observations. Journal of the Royal Statistical Society, series B, vol. 33, pp. 72-77.

Barraud, A.Y. (1977). A numerical algorithm to solve $A'XA-X=Q$. IEEE Transactions on Automatic Control, vol. AC-22, no. 5, pp. 883-885.

Basawa, I.V., Feign, P.D. and Heyde, C.C. (1976). Asymptotic properties of maximum likelihood estimators for stochastic processes. Sankhyā, series A, vol. 38, pp. 259-270.

Bhat, B.R. (1974). On the method of maximum likelihood for dependent observations. Journal of the Royal Statistical Society, series B, vol. 36, pp. 48-53.

Bowden, R. (1973). The theory of parametric identification. Econometrica, vol. 41, pp. 1069-1074.

Box, G.E.P. and Jenkins, G.M. (1976). Time Series Analysis, Forecasting and Control, revised edition. San Francisco: Holden Day.

Breusch, T.S. and Pagan, A.R. (1980). The Lagrange multiplier test and its applications to model specification in econometrics. Review of Economic Studies, vol. 47, pp. 239-254.

Caines, P.E. and Rissanen, J. (1974). Maximum likelihood estimation of parameters in multivariable Gaussian stochastic processes. IEEE Transactions on Information Theory, vol. IT-20. no. 1, pp. 102-104.

Caines, P.E. (1988). Linear Stochastic Systems. New York: John Wiley.

Chan, S.W., Goodwin, G.C. and Sin, K.S. (1984). Convergence properties of the Riccati difference equation in optimal filtering of nonstabilizable systems. IEEE Transactions on Automatic Control, vol. AC-29, no. 2, pp. 110-118.

Chow, G.C. (1975). Analysis and Control of Dynamic Economic Systems. New York: John Wiley.

Chow, G.C. (1981). Econometric Analysis by Control Methods. New York: John Wiley.

Chow, G.C. (1984). Random and changing coefficient models, in Z. Griliches and M.D. Intriligator (editors), Handbook of Econometrics, vol. II. Amsterdam: North-Holland.

Chow, G.C. and Lin A. (1976). Best linear unbiased estimation of missing observations in an economic time series. Journal of the American Statistical Association, vol. 71, pp. 719-721.

Cooley, T.F. and Prescott, E. (1976). Estimation in the presence of stochastic parameter variation. Econometrica, vol. 44, pp. 167-183.

Cramer, H. (1946). Mathematical Methods of Statistics. Princeton, N.J.: Princeton University Press.

Crowder, M. J. (1976). Maximum likelihood estimation for dependent observations. Journal of the Royal Statistical Society, series B, vol. 38, pp. 45-53.

Deistler, M. (1986). Linear dynamic errors in variables models, in J. Gani and M.B. Priestley (editors), Essays in Time Series and Allied Processes. Sheffield: Applied Probability Trust, Journal of Applied Probability, special volume 23A.

Engle, R.F. (1984). Wald, likelihood and Lagrange multiplier tests in econometrics, in Z. Griliches and M.D. Intriligator (editors), Handbook of Econometrics, vol. II. Amsterdam: North-Holland.

Engle, R.F. and Watson, M.W. (1981). A one factor multivariate time series model of metropolitan wage rates. Journal of the American Statistical Association, vol. 76, pp. 774-781.

Fisher, R. A. (1921). On the mathematical foundations of theoretical statistics. Philosophical Transactions of the Royal Society of London, series A, vol. 222, pp. 309-368.

Friedman, M. (1957). A Theory of the Consumption Function. Princeton, N.J.: Princeton University Press.

Gill, P., Murray, W. and Wright, M. (1981). Practical Optimization. New York: Academic Press.

Glover, K. and Willems, J.C. (1974). Parametrizations of linear dynamical systems: canonical forms and identifiability. IEEE Transactions on Automatic Control, vol. AC-19, no. 6, pp. 640-646.

Granger, C.W. and Morris, M.J. (1976). Time series modelling and interpretation. Journal of the Royal Statistical Society, series A, vol. 139, pp. 246-257.

Gupta, N.K. and Mehra, R.K. (1974). Computational aspects of maximum likelihood estimation and reduction in sensitivity function calculations. IEEE Transactions on Automatic Control, vol. AC-19, no.6, pp. 774-783.

Hamilton, J.D. (1985). Uncovering financial market expectations of inflation. Journal of Political Economy, vol. 93, pp. 1224-1241.

Hannan, E.J. (1980). The estimation of the order of an ARMA process. The Annals of Statistics, vol. 8, pp. 1071-1081.

Hannan, E.J. and Deistler, M. (1988). The Statistical Theory of Linear Systems. New York: John Wiley.

Harvey, A.C. (1981a). The Econometric Analysis of Time Series. Oxford: Philip Allan Publishers.

Harvey, A.C. (1981b). Time Series Models. Oxford: Philip Allan Publishers.

Harvey, A.C. and Tood, P.H.J. (1983). Forecasting economic time series with structural and Box-Jenkins models: A case study. Journal of Business and Economic Statistics, vol. 1, no. 4, pp. 299-315.

Harvey, A.C. and McKenzie, C.R. (1984). Missing observations in dynamic econometric models: A partial synthesis, in E. Parzen (editor), Time Series Analysis of Irregularly Observed Data. New York: Springer-Verlag.

Harvey, A.C. and Pierse, R.G. (1984). Estimating missing observations in economic time series. Journal of the American Statistical Association, vol. 79, pp. 125-131.

Hausman, J.A. and Watson, M.W. (1985). Errors in variables and seasonal adjustment procedures. Journal of the American Statistical Association, vol. 80, pp. 531-540.

Heijmans, R.D.H. and Magnus, J.R. (1986a). On the first order efficiency and asymptotic normality of maximum likelihood estimators obtained from dependent observations. Statistica Neerlandica, vol. 40, no. 3, pp. 169-187.

Heijmans, R.D.H. and Magnus, J.R. (1986b). Consistent maximum likelihood estimation with dependent observations: The general (non-normal) case and the normal case. Journal of Econometrics, vol. 32, pp. 253-285.

Hewer, G.A. (1971). An iterative technique for the computation of the steady state gains for the discrete optimal regulator. IEEE Transactions on Automatic Control, vol. AC-16, no. 4, pp. 382-384.

Jazwinsky, A.H. (1970). Stochastic Processes and Filtering Theory. New York: Academic Press.

Kalman, R.E. (1960). A new approach to linear filtering and prediction problems. Transactions of the ASME, Journal of Basic Engineering, series 82D, pp. 35-45.

Kang, H. (1981). Necessary and sufficient conditions for causality testing in multivariate time series. Journal of Time Series Analysis, vol 2, no. 2, pp. 95-101.

Kashyap,, R.L. (1977). A Bayesian comparison of different classes of dynamic models using empirical data. IEEE Transactions on Automatic Control, vol. AC-19, no. 5, pp. 715-727.

Kohn, R. and Ansley, C.F. (1983). Fixed interval estimation in state space models when some of the data are missing or aggregated. Biometrika, vol. 70, no. 3, pp. 683-688.

Ljung, L. (1987). Systems Identification, Theory for the User. Englewood Cliffs, N. J.: Prentice Hall.

Luenberger, D. (1984). Introduction to Linear and Nonlinear Programming, second edition. Reading, Massachusetts: Addison Wesley.

McDonald, J. and Darroch, J. (1983). Consistent estimation of equations with composite moving average disturbance terms. Journal of Econometrics, vol. 23, no. 2, pp. 253-267.

Mehra, R.K. (1976). Identification and estimation of the error-in-variables model (EVM) in structural form, in R.J.B. Web (editor), Stochastic Systems: Modelling Identification and Optimization, vol. 1, Amsterdam: North-Holland.

Morf, M., Sidhu, G.S. and Kailath, T. (1974). Some new algorithms for recursive estimation in constant, linear, discrete-time systems. IEEE Transactions on Automatic Control, vol. AC-19, no. 4, pp. 315-323.

Newbold, P. (1978). Feedback induced by measurement errors. International Economic Review, vol. 19, no. 3, pp. 787-791.

Newbold, P. (1983). Model checking in time series analysis, in A. Zellner (editor), Applied Time Series Analysis of Economic Data. U.S. Department of Commerce, Bureau of the Census.

Pagan, A. (1980). Some identification and estimation results for regression models with stochastically varying coefficients. Journal of Econometrics, vol. 13, no. 3, pp. 341-363.

Pappas, T., Laub, A.T. and Sandell, N.R. (1980). On the numerical solution of the algebraic Riccati equation. IEEE Transactions on Automatic Control, vol. AC-25, no. 4, pp. 631-641.

Parzen, E., editor (1984). Time Series Analysis of Irregularly Observed Data. New York: Springer-Verlag.

Quandt, R.E. (1983). Computational problems and methods, in Z. Griliches and M.D. Intriligator (editors), Handbook of Econometrics, vol. I. Amsterdam: North-Holland.

Rice, J.R. (1983). Matrix Computations and Mathematical Software. New York: MacGraw-Hill.

Rissanen, J. (1983). A universal prior for integers and estimation by minimum description lengths. The Annals of Statistics, vol. 8, pp. 416-431.

Rosenberg, B. (1973). The analysis of a cross section of time series by stochastically convergent parameter regression. Annals of Economic and Social Measurement, vol. 2, no. 4, pp. 399-428.

Rosenbrock, H.H. (1970). State-Space and Multivariable Theory. New York: John Wiley.

Rothenberg, T.J. (1971). Identification in parametric models. Econometrica, vol. 39, pp. 577-591.

Schweppe, F.C. (1965). Evaluation of likelihood functions for gaussian signals. IEEE Transactions of Information Theory, vol. IT-11, no. 1, pp. 61-70.

Silvey, S.D. (1961). A note on maximum likelihood in the case of dependent random variables. Journal of the Royal Statistical Society, series B, vol. 23, pp. 444-452.

Solo, V. (1986). Identifiability of time series models with errors in variables, in J. Gani and M.B. Priestley (editors), Essays in Time Series and Allied Processes. Sheffield: Applied Probability Trust, Journal of Applied Probability, special volume 23A.

Schwartz, G. (1978). Estimation the dimension of a model. The Annals of Statistics, vol. 6, pp. 461-464.

Terceiro, J. (1975). Error analysis of the linear stochastic control problem. Proceedings of the Eighth Hawaii International Conference on System Sciences. Western Periodical Company, pp. 66-78.

Wald, A. (1949). Note on the consistency of the maximum likelihood estimate, Annals of Mathematical Statistics, Vol. 20, pp. 595-601.

Watson, M.W. and Engle, R.F. (1983). Alternative algorithms for the estimation of dynamic factor, MIMIC and varying coefficient regression models. Journal of Econometrics, vol. 23, no. 3, pp. 385-400.

Wertz, V., Gevers, M. and Hannan, E.J. (1982). The determination of optimum structures for the state-space representation of multivariate stochastic processes. IEEE Transactions on Automatic Control, vol. AC-27, no. 6, pp. 1200-1210.

Zellner, A. and Palm, F. (1974). Time series analysis and simultaneous equation econometric models. Journal of Econometrics, vol. 2, no. 1, pp. 17-54.

AUTHOR INDEX

Aigner, D.J. 2, 4, 41, 56, 62
Akaike, H. 46, 47
Anderson, B.D.O. 2, 29, 36, 73
Ansley, C.F. 36, 49, 102, 103
Aoki, M. 2, 11

Bar-Shalom, Y. 42
Barraud, A.Y. 95
Basawa, I.V. 42
Bhat, B.R. 42
Bowden, R. 41
Box, G.E.P. 13, 45
Breusch, T.S. 45

Caines, P.E. 42, 43
Chan, S.W. 29, 32, 98
Chow, G.C. 13, 14, 53
Cooley, T.F. 13
Cramer, H. 41, 42
Crowder, M.J. 42

Darroch, J. 3
Deistler, M. 2, 43

Engle, R.F. 34, 45, 62, 86

Feign, P.D. 42
Fisher, R.A. 41
Friedman, M. 1

Gevers, M. 9
Gill, P., 38
Glover, K. 9
Goodwin, G.C. 29, 32, 98
Granger, C.W. 13
Gupta, N.K. 39

Hamilton, J.D. 3
Hannan, E.J. 9, 43, 47
Harvey, A.C. 14, 15, 26, 36, 49, 53, 93
Hausman, J.A. 3
Heijmans, R.D.H. 42
Hewer, G.A. 98, 99
Heyde, C.C. 42
Hsiao, C. 2, 4, 41, 56, 62

Jazwinsky, A.H. 73
Jenkins, G.M. 13, 45

Kailath, T. 74, 78
Kalman, R.E. 7
Kang, H. 19
Kapteyn, A. 2, 4, 41, 56, 62
Kashyap, R.L. 47
Kohn, R. 36, 49, 102, 103

Laub, A.T. 100
Lin A. 53
Ljung, L. 43
Luenberger, D. 37, 38

Magnus, J.R. 42
McDonald, J. 3
McKenzie, C.R. 53, 93
Mehra, R.K. 21, 39
Moore, J.B. 29, 36, 73
Morf, M. 74, 78
Morris, M.J. 13
Murray, P. 38

Newbold, P. 23, 45

Pagan, A.R. 42, 45
Palm, F. 18
Pappas, T. 100
Parzen, E. 49
Pierse, R.G. 49
Prescott, E. 13

Quandt, R.E. 38

Rice, J.R. 96
Rissanen, J. 42, 47
Rosenberg, B. 36, 93
Rosenbrock, H.H. 10
Rothenberg, T.J. 41

Sandell, N.R. 100
Schwartz, G. 47
Schweppe, F.C. 26
Sidhu, G.S. 74, 78
Silvey, S.D. 42
Sin, K.S. 29, 32, 98
Solo, V. 2

Terceiro, J. 93

Wald, A. 41, 42, 45
Wansbech, T. 2, 4, 41, 56, 62
Watson, M.W. 3, 34, 62, 86
Wertz, V. 9
Willems, J.C. 9
Wright, M. 38

Zellner, A. 18

SUBJECT INDEX

Akaike's Information Criterion 47
Asymptotic efficiency 41, 43
Asymptotic lower bound 33
Asymptotic normality 41-43
Asymptotic properties 25, 41-43
Asymptotic unbiasedness 42
Asymptotically efficient 2, 68
Asymptotically normal 2, 68
Autocorrelation function 10, 45
Autoregressive (AR) process 13, 28, 56, 58-61
Autoregressive-moving average (ARMA) process 2, 13, 53, 55

Canonical parametrization 9
Chandrasekhar equations 3, 16, 25, 28, 29, 37, 50, 57, 68, 70, 73, 74, 77
Cholesky factorization 25, 44
Coding theoretical ideas 46
Compactness hypothesis 43
Composite moving average disturbance terms 3, 69
Computational errors 73
Computational requirements 33, 38, 83, 84, 95
Conditional probability density 14, 26, 29
Consistency property 41
Consistent estimates 2, 45, 47, 68
Contemporaneous aggregation 49, 68, 101
Contemporaneous relationship 14, 15

Covariance equations 2, 88
Covariance matrix 2, 8, 12, 16, 20, 23, 25, 31, 33-35, 38, 41, 43-45, 50, 57, 70, 72, 86, 90, 92, 97
  singularity of 15
Cramer-Rao lower bound 43
Cross-correlation function 45

Data-based procedure 2
Determinist input 43
Distributed-lag functions 7
Dynamic matrix multipliers 6

Econometric model
  identification problem in 3, 8, 25
  improper nature of 5
  improper characteristic of 15
  reduced form of 5, 6, 10
  state-space form of 5, 24, 68
    minimal dimension of 10, 68, 77
  overparametrization of 2, 45
  structural form of 2, 5, 17, 19
  validation of 3
  verification of 23, 25, 45, 46, 62, 68
Efficiency 41, 43
Efficient parameter estimates 41
Endogenous variables 1, 2, 4, 5, 6, 7, 9, 17, 19, 20, 23, 49, 52, 53, 55, 56, 60, 63, 68, 69, 77, 101
Equally-spaced data 51

Equilibrium solutions 48
Errors in variables 1-3, 12, 17, 20, 21, 24, 25, 55, 56, 68, 77
Exact likelihood function 68
  analitycal expression 68
Exogenous variables 1, 2, 4, 5, 6, 7, 9, 12, 17, 19-21, 23, 45, 49, 52, 53, 55-58, 60-63, 68, 77, 101

Final form 6, 7, 9, 10
Fixed-interval smoothing algorithm 51, 53, 101, 104
Flow variables 52

General Criterion for Structural Selection 46
Gradient 3, 25, 31, 35, 46, 62, 80, 84
  adimensional expression for 25, 43
  exact expressions for 46
Gradient methods 37

Hessian 3, 25, 31, 33, 35, 37, 38, 46, 47, 62, 83
  adimensional expression for 25, 43
  calculation of 46, 38, 83
Householder transformations 95

Identifiable formulations 8, 9, 40, 41, 58
Identifiability 2, 42, 51
Identification of the model 2, 3, 8, 25, 41
Improper model 14, 15
Information matrix 2, 3, 25, 33-35, 38-40, 45, 46, 61, 62, 68, 85
  eigenvalues and eigenvectors of 25, 39, 40, 61

Information matrix (Cont.)
  exact expression of 3, 34, 35, 45, 46, 61, 62, 68
  exact analytical expression of 62
  calculation of 85
Information filter 36
Initial conditions 3, 16, 25, 29, 35-37, 62, 68, 74-77, 89-91, 93, 97, 103
  calculation of 90
Innovation process 25, 27, 28, 35, 38, 45, 90
Input-output equivalent state-space forms 46
Input vectors 8
Instantaneous coupling 6
Interpolation method 53

Joint probability density function 25

Kalman filter 3, 16, 25-31, 33, 35-37, 50, 51, 68, 70-73, 77, 93, 97, 102
  information forms of 77
  initial conditions for 29, 35-37, 97
  innovation process of 25, 35, 45
  stability of 29
Kronecker product 95

Lagrange-multiplier test 26, 45
Likelihood function 2, 3, 16, 25, 26, 28, 29, 31, 33, 35, 36, 41, 68, 80, 97
  gradient of 3, 25, 31, 35, 80, 97
  analytical expression of 3, 25, 35
  calculation of 31, 80, 97

Likelihood function (Cont.)
  hessian of 3, 25, 31, 33, 35, 83
    analytical expression of 3, 25, 35
    calculation of 31, 33, 83
  maximization of 16, 29
Likelihood-ratio test 45
Locally identifiable 41
Log-likelihood function 41
  first derivative information of 29
  second derivative information of 29
Long-run equilibrium 48
Long term behaviour 16
Lower bound 33, 43, 45
Lyapunov equation 36, 76, 89, 94, 96, 97, 99, 100
  solution of 36, 76, 97, 100
Lyapunov algebraic equation 32, 3, 89, 95

Markov process 14
Mathematical program with restrictions 28
Matrix
  adjoint 18
  determinant 18
  eigenvalues 25, 38-41, 61, 95, 98, 100
  eigenvectors 25, 39, 40, 61
  generalized inverse 93, 102
  generalized eigenvalues and eigenvectors 100
  orthogonal 95
  positive definite 7, 38, 44, 92
  positive semidefinite 7, 25, 38, 44, 73, 76, 98, 99
  Schur form 95-97
  singular 6, 36, 38, 39, 100
  upper Hessenberg form 95

Maximum likelihood 2, 41, 46
Maximum likelihood estimation 2, 24, 25, 41-43, 46, 68, 69, 75, 90, 91
Measurement error 1, 2, 3, 4, 17, 20-25, 41, 49, 52, 53, 56, 62, 63, 68, 69, 77
  correlated measurement errors 3, 53, 68
Method of "scoring" 38
Minimal dimension 2, 14-16
Minimal realization 10
Minimum Description Length 47
Minimum dimension 5
Missing observations 3, 4, 49-51, 53, 62, 63, 68, 69, 101
Monte-Carlo analysis 61
Multivariate ARMA model 2, 19, 23, 53, 55

Negative log-likelihood function 27, 75, 80
Newton-Raphson method 38
Noise term 12, 13, 15, 73
Non-identifiable parameters 40
Nonminimal dimensional model 14
Nonstationary model 29, 36, 76, 77
Normality 26, 41-43
Null hypothesis 45
Numerical optimization 3, 29
Numerically efficient procedures 37

Observability condition 10, 11
Observation equation 8, 14, 15, 21, 23, 73, 101
Observation noise, errors 1, 15, 17, 23, 41, 45, 49, 50, 53, 55, 60-62
  correlated observation errors 15, 41, 49

One-dimensional search procedure 37
Optimization algorithm 3, 25, 62
Optimization methods 29, 37, 38
Output vector 8
Overparametrization 1, 2, 45

Parsimonious formulation 13
Partial adjustment 56
Partial correlation function 45
Permanent income hypothesis 1
Probability density function 12, 15
Proper formulations 14, 15

Quarterly variables 51, 53

Random coefficient 13, 14
Rational expectations 3, 69
Rational functions 6
Reachability conditions 11
Reduced form 5, 6, 9, 10, 11, 45
Regularity condition 1, 42
Restrictions 2, 8, 9, 16, 18, 23, 28, 29, 40, 41, 46, 48
Riccati matrix equation 3, 25, 28, 57, 68, 75, 77, 94, 97, 98, 100
Riccati algebraic matrix equation 32

Sampling interval 14
Seasonal adjustment of time series 3, 69
Sensitivity equations 30, 31, 97
Sensitivity analysis 93
Shift operator 6
Short-run behaviour 48
Signal-to-noise ratio 58-61
Similarity transformation 8
Small samples 16
Spurious feedback 1, 23
Stability conditions 16

State-space model, form, formulation 1, 2, 14, 42, 46, 48, 58, 68, 71, 77
State transition equation 8, 15
State vector, variables 2, 3, 5, 7-12, 14-16, 25, 28, 51, 73, 92, 104
  minimal dimension for 2, 15, 16
Stationary model 29, 36, 76, 77
Steady-state 7, 32, 48
  expression 32
  matrix multiplier 7, 48
  values 32, 33
Stochastic input 43
Stochastic realization problem 10
Stock variables 53
Storage requirement 32
System noise 15

Temporal aggregates 3, 49, 53, 68
Temporal aggregation 51, 53, 101
Theoretical information 46
Time series model 3, 28, 69
Time-varying system 28

Uncertainty hiperellipsoid 40
Unequally-spaced data 51
Unit roots 29

Wald test 45
White noise process 7, 9, 12, 17, 19, 20, 45, 53, 57-60, 70

Vol. 236: G. Gandolfo, P.C. Padoan, A Disequilibrium Model of Real and Financial Accumulation in an Open Economy. VI, 172 pages. 1984.

Vol. 237: Misspecification Analysis. Proceedings, 1983. Edited by T.K. Dijkstra. V, 129 pages. 1984.

Vol. 238: W. Domschke, A. Drexl, Location and Layout Planning. IV, 134 pages. 1985.

Vol. 239: Microeconomic Models of Housing Markets. Edited by K. Stahl. VII, 197 pages. 1985.

Vol. 240: Contributions to Operations Research. Proceedings, 1984. Edited by K. Neumann and D. Pallaschke. V, 190 pages. 1985.

Vol. 241: U. Wittmann, Das Konzept rationaler Preiserwartungen. XI, 310 Seiten. 1985.

Vol. 242: Decision Making with Multiple Objectives. Proceedings, 1984. Edited by Y.Y. Haimes and V. Chankong. XI, 571 pages. 1985.

Vol. 243: Integer Programming and Related Areas. A Classified Bibliography 1981–1984. Edited by R. von Randow. XX, 386 pages. 1985.

Vol. 244: Advances in Equilibrium Theory. Proceedings, 1984. Edited by C.D. Aliprantis, O. Burkinshaw and N.J. Rothman. II, 235 pages. 1985.

Vol. 245: J.E.M. Wilhelm, Arbitrage Theory. VII, 114 pages. 1985.

Vol. 246: P.W. Otter, Dynamic Feature Space Modelling, Filtering and Self-Tuning Control of Stochastic Systems. XIV, 177 pages. 1985.

Vol. 247: Optimization and Discrete Choice in Urban Systems. Proceedings, 1983. Edited by B.G. Hutchinson, P. Nijkamp and M. Batty. VI, 371 pages. 1985.

Vol. 248: Plural Rationality and Interactive Decision Processes. Proceedings, 1984. Edited by M. Grauer, M. Thompson and A.P. Wierzbicki. VI, 354 pages. 1985.

Vol. 249: Spatial Price Equilibrium: Advances in Theory, Computation and Application. Proceedings, 1984. Edited by P.T. Harker. VII, 277 pages. 1985.

Vol. 250: M. Roubens, Ph. Vincke, Preference Modelling. VIII, 94 pages. 1985.

Vol. 251: Input-Output Modeling. Proceedings, 1984. Edited by A. Smyshlyaev. VI, 261 pages. 1985.

Vol. 252: A. Birolini, On the Use of Stochastic Processes in Modeling Reliability Problems. VI, 105 pages. 1985.

Vol. 253: C. Withagen, Economic Theory and International Trade in Natural Exhaustible Resources. VI, 172 pages. 1985.

Vol. 254: S. Müller, Arbitrage Pricing of Contingent Claims. VIII, 151 pages. 1985.

Vol. 255: Nondifferentiable Optimization: Motivations and Applications. Proceedings, 1984. Edited by V.F. Demyanov and D. Pallaschke. VI, 350 pages. 1985.

Vol. 256: Convexity and Duality in Optimization. Proceedings, 1984. Edited by J. Ponstein. V, 142 pages. 1985.

Vol. 257: Dynamics of Macrosystems. Proceedings, 1984. Edited by J.-P. Aubin, D. Saari and K. Sigmund. VI, 280 pages. 1985.

Vol. 258: H. Funke, Eine allgemeine Theorie der Polypol- und Oligopolpreisbildung. III, 237 pages. 1985.

Vol. 259: Infinite Programming. Proceedings, 1984. Edited by E.J. Anderson and A.B. Philpott. XIV, 244 pages. 1985.

Vol. 260: H.-J. Kruse, Degeneracy Graphs and the Neighbourhood Problem. VIII, 128 pages. 1986.

Vol. 261: Th.R. Gulledge, Jr., N.K. Womer, The Economics of Made-to-Order Production. VI, 134 pages. 1986.

Vol. 262: H.U. Buhl, A Neo-Classical Theory of Distribution and Wealth. V, 146 pages. 1986.

Vol. 263: M. Schäfer, Resource Extraction and Market Structure. XI, 154 pages. 1986.

Vol. 264: Models of Economic Dynamics. Proceedings, 1983. Edited by H.F. Sonnenschein. VII, 212 pages. 1986.

Vol. 265: Dynamic Games and Applications in Economics. Edited by T. Başar. IX, 288 pages. 1986.

Vol. 266: Multi-Stage Production Planning and Inventory Control. Edited by S. Axsäter, Ch. Schneeweiss and E. Silver. V, 264 pages. 1986.

Vol. 267: R. Bemelmans, The Capacity Aspect of Inventories. IX, 165 pages. 1986.

Vol. 268: V. Firchau, Information Evaluation in Capital Markets. VII, 103 pages. 1986.

Vol. 269: A. Borglin, H. Keiding, Optimality in Infinite Horizon Economies. VI, 180 pages. 1986.

Vol. 270: Technological Change, Employment and Spatial Dynamics. Proceedings 1985. Edited by P. Nijkamp. VII, 466 pages. 1986.

Vol. 271: C. Hildreth, The Cowles Commission in Chicago, 1939–1955. V, 176 pages. 1986.

Vol. 272: G. Clemenz, Credit Markets with Asymmetric Information. VIII, 212 pages. 1986.

Vol. 273: Large-Scale Modelling and Interactive Decision Analysis. Proceedings, 1985. Edited by G. Fandel, M. Grauer, A. Kurzhanski and A.P. Wierzbicki. VII, 363 pages. 1986.

Vol. 274: W.K. Klein Haneveld, Duality in Stochastic Linear and Dynamic Programming. VII, 295 pages. 1986.

Vol. 275: Competition, Instability, and Nonlinear Cycles. Proceedings, 1985. Edited by W. Semmler. XII, 340 pages. 1986.

Vol. 276: M.R. Baye, D.A. Black, Consumer Behavior, Cost of Living Measures, and the Income Tax. VII, 119 pages. 1986.

Vol. 277: Studies in Austrian Capital Theory, Investment and Time. Edited by M. Faber. VI, 317 pages. 1986.

Vol. 278: W.E. Diewert, The Measurement of the Economic Benefits of Infrastructure Services. V, 202 pages. 1986.

Vol. 279: H.-J. Büttler, G. Frei and B. Schips, Estimation of Disequilibrium Models. VI, 114 pages. 1986.

Vol. 280: H.T. Lau, Combinatorial Heuristic Algorithms with FORTRAN. VII, 126 pages. 1986.

Vol. 281: Ch.-L. Hwang, M.-J. Lin, Group Decision Making under Multiple Criteria. XI, 400 pages. 1987.

Vol. 282: K. Schittkowski, More Test Examples for Nonlinear Programming Codes. V, 261 pages. 1987.

Vol. 283: G. Gabisch, H.-W. Lorenz, Business Cycle Theory. VII, 229 pages. 1987.

Vol. 284: H. Lütkepohl, Forecasting Aggregated Vector ARMA Processes. X, 323 pages. 1987.

Vol. 285: Toward Interactive and Intelligent Decision Support Systems. Volume 1. Proceedings, 1986. Edited by Y. Sawaragi, K. Inoue and H. Nakayama. XII, 445 pages. 1987.

Vol. 286: Toward Interactive and Intelligent Decision Support Systems. Volume 2. Proceedings, 1986. Edited by Y. Sawaragi, K. Inoue and H. Nakayama. XII, 450 pages. 1987.

Vol. 287: Dynamical Systems. Proceedings, 1985. Edited by A.B. Kurzhanski and K. Sigmund. VI, 215 pages. 1987.

Vol. 288: G.D. Rudebusch, The Estimation of Macroeconomic Disequilibrium Models with Regime Classification Information. VII, 128 pages. 1987.

Vol. 289: B.R. Meijboom, Planning in Decentralized Firms. X, 168 pages. 1987.

Vol. 290: D.A. Carlson, A. Haurie, Infinite Horizon Optimal Control. XI, 254 pages. 1987.

Vol. 291: N. Takahashi, Design of Adaptive Organizations. VI, 140 pages. 1987.

Vol. 292: I. Tchijov, L. Tomaszewicz (Eds.), Input-Output Modeling. Proceedings, 1985. VI, 195 pages. 1987.

Vol. 293: D. Batten, J. Casti, B. Johansson (Eds.), Economic Evolution and Structural Adjustment. Proceedings, 1985. VI, 382 pages. 1987.

Vol. 294: J. Jahn, W. Krabs (Eds.), Recent Advances and Historical Development of Vector Optimization. VII, 405 pages. 1987.

Vol. 295: H. Meister, The Purification Problem for Constrained Games with Incomplete Information. X, 127 pages. 1987.

Vol. 296: A. Börsch-Supan, Econometric Analysis of Discrete Choice. VIII, 211 pages. 1987.

Vol. 297: V. Fedorov, H. Läuter (Eds.), Model-Oriented Data Analysis. Proceedings, 1987. VI, 239 pages. 1988.

Vol. 298: S.H. Chew, Q. Zheng, Integral Global Optimization. VII, 179 pages. 1988.

Vol. 299: K. Marti, Descent Directions and Efficient Solutions in Discretely Distributed Stochastic Programs. XIV, 178 pages. 1988.

Vol. 300: U. Derigs, Programming in Networks and Graphs. XI, 315 pages. 1988.

Vol. 301: J. Kacprzyk, M. Roubens (Eds.), Non-Conventional Preference Relations in Decision Making. VII, 155 pages. 1988.

Vol. 302: H.A. Eiselt, G. Pederzoli (Eds.), Advances in Optimization and Control. Proceedings, 1986. VIII, 372 pages. 1988.

Vol. 303: F.X. Diebold, Empirical Modeling of Exchange Rate Dynamics. VII, 143 pages. 1988.

Vol. 304: A. Kurzhanski, K. Neumann, D. Pallaschke (Eds.), Optimization, Parallel Processing and Applications. Proceedings, 1987. VI, 292 pages. 1988.

Vol. 305: G.-J.C.Th. van Schijndel, Dynamic Firm and Investor Behaviour under Progressive Personal Taxation. X, 215 pages. 1988.

Vol. 306: Ch. Klein, A Static Microeconomic Model of Pure Competition. VIII, 139 pages. 1988.

Vol. 307: T.K. Dijkstra (Ed.), On Model Uncertainty and its Statistical Implications. VII, 138 pages. 1988.

Vol. 308: J.R. Daduna, A. Wren (Eds.), Computer-Aided Transit Scheduling. VIII, 339 pages. 1988.

Vol. 309: G. Ricci, K. Velupillai (Eds.), Growth Cycles and Multisectoral Economics: the Goodwin Tradition. III, 126 pages. 1988.

Vol. 310: J. Kacprzyk, M. Fedrizzi (Eds.), Combining Fuzzy Imprecision with Probabilistic Uncertainty in Decision Making. IX, 399 pages. 1988.

Vol. 311: R. Färe, Fundamentals of Production Theory. IX, 163 pages. 1988.

Vol. 312: J. Krishnakumar, Estimation of Simultaneous Equation Models with Error Components Structure. X, 357 pages. 1988.

Vol. 313: W. Jammernegg, Sequential Binary Investment Decisions. VI, 156 pages. 1988.

Vol. 314: R. Tietz, W. Albers, R. Selten (Eds.), Bounded Rational Behavior in Experimental Games and Markets. VI, 368 pages. 1988.

Vol. 315: I. Orishimo, G.J.D. Hewings, P. Nijkamp (Eds.), Information Technology: Social and Spatial Perspectives. Proceedings, 1986. VI, 268 pages. 1988.

Vol. 316: R.L. Basmann, D.J. Slottje, K. Hayes, J.D. Johnson, D.J. Molina, The Generalized Fechner-Thurstone Direct Utility Function and Some of its Uses. VIII, 159 pages. 1988.

Vol. 317: L. Bianco, A. La Bella (Eds.), Freight Transport Planning and Logistics. Proceedings, 1987. X, 568 pages. 1988.

Vol. 318: T. Doup, Simplicial Algorithms on the Simplotope. VIII, 262 pages. 1988.

Vol. 319: D.T. Luc, Theory of Vector Optimization. VIII, 173 pages. 1989.

Vol. 320: D. van der Wijst, Financial Structure in Small Business. VII, 181 pages. 1989.

Vol. 321: M. Di Matteo, R.M. Goodwin, A. Vercelli (Eds.), Technological and Social Factors in Long Term Fluctuations. Proceedings. IX, 442 pages. 1989.

Vol. 322: T. Kollintzas (Ed.), The Rational Expectations Equilibrium Inventory Model. XI, 269 pages. 1989.

Vol. 323: M.B.M. de Koster, Capacity Oriented Analysis and Design of Production Systems. XII, 245 pages. 1989.

Vol. 324: I.M. Bomze, B.M. Pötscher, Game Theoretical Foundations of Evolutionary Stability. VI, 145 pages. 1989.

Vol. 325: P. Ferri, E. Greenberg, The Labor Market and Business Cycle Theories. X, 183 pages. 1989.

Vol. 326: Ch. Sauer, Alternative Theories of Output, Unemployment, and Inflation in Germany: 1960–1985. XIII, 206 pages. 1989.

Vol. 327: M. Tawada, Production Structure and International Trade. V, 132 pages. 1989.

Vol. 328: W. Güth, B. Kalkofen, Unique Solutions for Strategic Games. VII, 200 pages. 1989.

Vol. 329: G. Tillmann, Equity, Incentives, and Taxation. VI, 132 pages. 1989.

Vol. 330: P.M. Kort, Optimal Dynamic Investment Policies of a Value Maximizing Firm. VII, 185 pages. 1989.

Vol. 331: A. Lewandowski, A.P. Wierzbicki (Eds.), Aspiration Based Decision Support Systems. X, 400 pages. 1989.

Vol. 332: T.R. Gulledge, Jr., L.A. Litteral (Eds.), Cost Analysis Applications of Economics and Operations Research. Proceedings. VII, 422 pages. 1989.

Vol. 333: N. Dellaert, Production to Order. VII, 158 pages. 1989.

Vol. 334: H.-W. Lorenz, Nonlinear Dynamical Economics and Chaotic Motion. XI, 248 pages. 1989.

Vol. 335: A.G. Lockett, G. Islei (Eds.), Improving Decision Making in Organisations. Proceedings. IX, 606 pages. 1989.

Vol. 336: T. Puu, Nonlinear Economic Dynamics. VII, 119 pages. 1989.

Vol. 337: A. Lewandowski, I. Stanchev (Eds.), Methodology and Software for Interactive Decision Support. VIII, 309 pages. 1989.

Vol. 338: J.K. Ho, R.P. Sundarraj, DECOMP: an Implementation of Dantzig-Wolfe Decomposition for Linear Programming. VI, 206 pages. 1989.

Vol. 339: J. Terceiro Lomba, Estimation of Dynamic Econometric Models with Errors in Variables. VIII, 116 pages. 1990.